Lab Reports and Science Books

Lucy Calkins, Lauren Kolbeck, and Monique Knight

Photography by Peter Cunningham

HEINEMANN ◆ PORTSMOUTH, NH

This book is dedicated to Phil, who encourages and inspires me each and every day. —Lauren

This book is dedicated to Tom Corcoran, whose knowledge of science teaching enriches this book and my life. —Lucy

This book is dedicated to my family, whose love and support carry me. —Monique

DEDICATED TO TEACHERS™

*first*hand
An imprint of Heinemann
361 Hanover Street
Portsmouth, NH 03801–3912
www.heinemann.com

Offices and agents throughout the world

The authors and publisher wish to thank those who have generously given permission to reprint borrowed material:

Excerpt from *Forces and Motion (Hands-on Science)* by John Graham, illustrated by David Le Jars. Copyright © 2013 Kingfisher Publications Plc. Reprinted by permission of Kingfisher, an imprint of Macmillan Children's Books, London, UK.

Excerpts from http://www.ehow.com/info_8580833_motion-force-kids.html. Article courtesy of eHow.com and author Asa Jomard.

Cataloging-in-Publication data is on file with the Library of Congress.

ISBN-13: 978-0-325-04729-4
ISBN-10: 0-325-04729-4

Production: Elizabeth Valway, David Stirling, and Abigail Heim
Cover and interior designs: Jenny Jensen Greenleaf
Series includes photographs by Peter Cunningham, Nadine Baldasare, and Elizabeth Dunford
Composition: Publishers' Design and Production Services, Inc.
Manufacturing: Steve Bernier

Printed in the United States of America on acid-free paper
17 16 15 14 13 VP 3 4 5

Acknowledgments

THIS WAS AN EXCITING UNIT TO WRITE, one that involved a lot of new thinking and therefore a lot of collaborators. We are grateful to the people who have made us wiser about science education, starting with Tom Corcoran, who directs the Center for Policy Research at Teachers College (CPRE) and is engaged in science education across the world, often in collaboration with Merck Institute for Science Education. Over the past five years, colleagues at the Teachers College Reading and Writing Project have joined us in developing what we hope are state-of-the-art ideas about content literacy, and we are especially grateful to Kathleen Tolan and Amanda Hartman for that work. The methods in this book represent not only our best knowledge about content literacy and about science, but also our best knowledge about teaching writing, and for that our list of thank-yous is long indeed. The entire body of work at the Project stands on the shoulders of two men who have both left us over the past few years: Donald Murray and Donald Graves. They would have loved to imagine young scientists rolling trucks down ramps and racing to record their findings!

Closer to home, our ideas benefit from the thinking done at the Project's famous Thursdays. For a full day once a week, we gather with the others who join us as part of the Teachers College Reading and Writing Project team to take our best ideas about teaching reading, writing, and content literacy and to make those ideas better. On these Thursdays and in the Project, a few people especially sparked new thinking about primary writing, and we are grateful to Celena Larkey, Elizabeth Dunford, Shanna Schwartz, Laurie Pessah, and Lindsay Mann.

This book itself was written not only by the three of us, but by others who shared the pen—borrowing on a tradition in primary classrooms. Kate Montgomery, especially, joined our team, added layers of insight and knowledge, and helped us keep the book unified and cohesive.

Kim Thompson was always ready, asking, "What can I do to help?" Jasmine Junsay was a research assistant for the entire project, working with children and teachers to help pioneer every new twist and turn. Editor Zoe Ryder White demonstrated the real power of close reading, for she saw not only what was on the page but also what was almost but not quite on the page.

Finally, we are grateful to all the many teachers and children who have been our first and most important mentors. We are especially grateful to Sharon Fine and Irene Quiles of PS 50, Staten Island, and to Giselle Gault, Katie Dellostritto, and Lindsay Wilkes of PS 58 in Brooklyn.

The class described in this unit is a composite class, with children and partnerships of children gleaned from classrooms in very different contexts, then put together here. We wrote the units this way to bring you both a wide array of wonderful, quirky, various children and also to illustrate for you the predictable (and unpredictable) situations and responses this unit has created in classrooms across the nation and world.

—Lucy, Lauren, and Monique

Contents

Welcome to the Unit

THIS UNIT IS ONE OF THE MOST GROUNDBREAKING IN THE SERIES. The aim is to teach students more about information writing and, specifically, about the kinds of information writing that scientists are apt to do. To teach this, science instruction is folded into the writing workshop. The structures and methods of a writing workshop remain consistent, but during minilessons and small groups students are taught not only about writing but also about force and motion and about the scientific method. During the workshop time, children often divide their time between engaging in experiments and writing for the purposes that scientists write.

As part of this, children will send cars zooming down homemade ramps, hypothesize and record findings, speculate to develop theories and organize further experiments, and write lab reports. By inviting young people to write like scientists, the unit illuminates the work of being an information writer. The unit also helps students know more about information texts. Like many of this kind of text, lab reports are a conglomerate. At least a portion of a lab report is a how-to or procedural text. In another portion of the lab report, the scientist uses charts and graphs to synthesize data. Of course, all of the writing in a science inquiry is not writing that reaches readers. Children write to learn more as well as to teach.

During the final portion of the unit, young scientists will write information books in which they aim to teach others what they have just learned. The entire unit, then, conveys a way writing can be used to accelerate and reveal learning.

This unit is important for many reasons. First, it is aligned with many of the expectations for second-graders, and even third-graders, as outlined in the Common Core State Standards. Throughout the unit, children will "write informative/explanatory texts in which they introduce a topic, use facts and definitions to develop points, and provide a concluding statement or section" (W.2.2.). Both on their own and with guidance and support of one another and you, their teacher, they will "focus on a topic and strengthen writing as needed by revising and editing" (W.2.5) and "participate in shared research and writing projects (e.g., read a number of books on a single topic to produce a report; record science observations)" (W.2.7). Finally, conventions continue to play a role in this portion of the second-grade curriculum. As children write scientific reports, you will teach them to focus on capitalization, punctuation, and spelling (L.2.2).

To us, the unit is an especially important one because information writing is, after all, the type of writing that is nearest and dearest to our hearts. This entire series is a gigantic information writing text, and the character of this project illustrates the character of the genre. Notice, for example, the way information writing amplifies and extends our discipline—teaching. That's one of the unique characteristics of information writing. It takes on the character of the subject. If you are an economist and you write information texts about your field, your writing enables you to see patterns in your data and to inform others about what you see. If you are a political leader and you write about your leadership, your writing enables you to reflect upon and commit yourself to principles that underlie actions. That is, whenever writing is embedded in a discipline, the writer is writing to teach others about his or her discipline and also writing to learn more.

It seemed important to us that in this series, at least half of the information writing that children do is embedded in discipline-based studies. In fourth and fifth grade, those information writing units are part of history units, and in this second-grade unit, we have linked information writing and science. Although we embedded this in a unit on force and motion, the instruction could be transferred and applied to another unit or even another field.

We chose to focus on science because we want to champion the discipline. We know that learning about the world—about rocks and clouds, wheels and wires, is part and parcel of what childhood should be about. We know that hands-on, active learning is part and parcel of childhood, too, and wanted to show how easy it is to combine writing instruction and that sort of learning. And we know that boys especially love science and that a unit such as this gives boys another way to connect with writing and to grasp the power of writing. They will develop a hypothesis, test out theories, record trials, and explain results and conclusions. As children live like scientists, they will study data, noticing patterns and creating models to represent the data.

The decision to embed the unit not just in science but in the physical sciences and especially in forces and motion was made because many science educators believe this is the most overlooked and critical branch of science. Also, we wanted to give students opportunities to cycle through the scientific method, and we are aware that experiments involving wheels, ramps, and small machines tend to produce findings in a matter of minutes—and it can take months before a plant's leaves grow yellow or an animal learns to run differently through the maze.

You may launch the unit with some apprehension, thinking that you hadn't intended for your decision to teach writing to necessarily involve you in teaching science. This is our promise to you: although yes, the unit will be a bit unusual and may require you to devote a bit of extra time to understanding the expectations of the unit, it will tap such profound levels of motivation in your students that you won't end up worrying about the little confusions. Something big will get started in this unit, and you'll feel it. By channeling your students' burgeoning interest in science into the writing workshop, you'll be showing learners that writing need not be an end in and of itself but that it can also be a tool for learning.

OVERVIEW OF THE UNIT

In the first bend of this unit, students will study a shared class science topic, which is unusual because usually in a writing workshop, students pursue topics of their own choosing and their instruction focuses on writing well, not on the content. But the first day's workshop is even more unusual because you won't be teaching the kids a small, easily replicable strategy that they can quickly learn to do independently. Instead, today's session puts children through the more complicated sequence that spans an entire experiment—starting with asking and recording a question, then designing and finally conducting multiple trials of a simple experiment. Children jot and sketch as they go, getting a four-page lab report booklet started in the meeting area, with their hypotheses on one page, their procedures on another, their results on a third, and their conclusions on a fourth. Later, you'll help students reflect on this process, but for now, it is enough to move through the entire process (and to write about all steps within the process) all within one writing workshop.

In the second bend of the unit, "Writing to Teach Others about Our Discoveries," your goal will be to help your students begin to internalize the scientific procedures and writing processes they encountered in Bend I so they can teach others. You'll ignite students' enthusiasm for the new round of investigation by reminding them that they need to become real scientists, and to do so they should join the scientific community of their school. As such—you might tell them—it will be essential that they communicate clearly all that they have learned. In this session, you will channel students toward writing to teach others about their discoveries. You'll emphasize the importance of writing precise procedures so their experiments can be replicated. You'll introduce mentor texts for students not only so that they can see how real-world lab reports go but also so that they might revisit and improve lab reports already in progress.

By the end of this bend, your students should feel like they are expert enough at writing lab reports that they can do so without scaffolding. Students will be able to independently design and conduct an experiment, recording their processes on the lab reports they construct as they progress through the work. They'll meanwhile learn ways to lift the level of their work. They'll learn to write with domain-specific vocabulary and will do so not only in new writing but also will reread the work they've already written, revising this aspect of that work. They'll learn, too, about ways to elaborate and, again, will use those ways both to write new lab reports and to revise previously written ones.

The third and final bend of the unit marks an important turning point. You'll invite students to write an information book that teaches readers all about a topic that the writer knows well and that—here's the trick—relates in some ways to the research children have just done on forces and motion. Imagine you were asked to do this. You've learned something about Newton's first law—about inertia and friction and gravity and simple machines. Although you've done some experiments to get a handle on that knowledge, it is still complicated for you. Now you are challenged to take a topic that you know well related in some way to forces and motion and to teach that subject to readers. So, I know dogs well. I could write an information book about dogs. But bringing in information about forces and motion—all of a sudden the job becomes very complicated indeed!

Children will, of course, get some assistance. For example, whether they write about bicycling or golf or skateboarding or skating, a good deal of what they can say about forces and motion will be similar. You'll help children apply their knowledge to these subjects, and you'll help them learn from each other's work.

Meanwhile, a good deal of your teaching will help children with the special challenges of this sort of information writing. For example, the drawings in a science information text tend to be different from those in a history text. The former are more apt to contain a sequence of numbers or arrows or split screens. It is helpful if children know this and if they try it themselves. Comparisons have a special place in science books. To model how to make this kind of writing, you'll rely on a mentor text. In the first bend of the unit we recommended John Graham's *Hands-On Science: Forces and Motion.* In Bend III, we recommended Stephen Biesty's *Incredible Cross Sections* to study during the minilesson. Of course, children will be encouraged not only to read the texts that you highlight but also to find their own mentor texts. You'll help students read texts closely, studying techniques the authors have used and thinking about the reasons the authors made the choices they did. As part of this, students will notice the use of domain-specific vocabulary in information texts. This close analytical reading work reflects the craft and structure standards of the Common Core State Standards for Reading Informational Text (2.4, 2.5, 2.6), and it ties reading and writing workshop tightly together.

ASSESSMENT

You will probably have assessed your students' information writing by conducting an on-demand assessment at the start of the school year. If this is your second unit of study, you may decide not to conduct a second assessment just yet, postponing that second on-demand until the end of this unit. Of course, if you neglected to assess your students' information writing skills at the start of the year, you will absolutely want to do so now.

In *Writing Pathways: Performance Assessments and Learning Progressions, K–5,* we lay out the details of the on-demand assessment, as well as provide you with a specific prompt for assessing your children's information writing. We hope you follow those guidelines so that the conditions under which you conduct the assessment are the same as those in other schools across the nation. But the really important thing is that you and your colleagues determine a single way in which all of you, as second-grade teachers, will administer the assessment. It is essential that the conditions are as similar across classrooms as you can manage, so that you can compare the results. The prompt is:

> "Think of a topic that you've studied or that you know a lot about. Tomorrow, you will have forty-five minutes to write an informational (or all-about) text that teaches others interesting and important information and ideas about that topic. If you want to find and use information from a book or another outside source to help you with this writing, you may bring that with you tomorrow. Please keep in mind that you'll have only forty-five minutes to complete this. You will have only this one period, so you'll need to plan, draft, revise, and edit in one sitting. Write in a way that shows all that you know about information writing."

You'll want to let your youngsters know that they can use the whole writing workshop (forty-five minutes) to write the best that they can and to fill their pages with as much information as they can teach. Because you will examine the work that your children produce as part of a K–5 learning progression, we recommend that you give them some additional instructions for this task.

"In your writing, make sure you:

- introduce the topic you will teach about
- include lots of information
- organize your writing
- use transition words
- write an ending."

This on-demand work will give you a snapshot of your students' strengths and needs. Assuming you are teaching this unit in the fall of second grade, you should expect that your students' writing (if they come from a strong writing curriculum in first grade) will be roughly at the first-grade level at the start of second grade. That is, remember that children often slip backward in the summer, so if your youngsters are achieving end-of-first-grade standards in October of second grade, that means they are positioned for the year ahead.

Of course, you will undoubtedly see that children have strengths in some areas and needs in others. Pay attention to these, design small groups from these data, and use this information to help maximize progress during the upcoming unit. To plan ways to tailor your instruction to your students, you will probably want to think with your grade level team about what the results of the writing assessment show you. If all the teachers at your grade level bring examples of writing that you feel represent levels 1, 2, 3, and 4 (according to the rubric) to a study group, you can create your own set of exemplar pieces as a grade team. You can then use this set of exemplars to quickly assess the rest of your students' work, although assigning an exact level for each student is not as important as the conversations you have about the work, which will enable you to align your vision as a grade. All of this information will help you plan predictable small groups for your advanced and struggling writers, as well as next steps for teaching.

You can use the checklists, which are aligned to the learning progressions, to help your students self-assess, set goals, and develop action plans. Students will be able to see for themselves what skills they already know how to do as writers that they'll carry into this unit, as well as some skills toward which to reach. The descriptors will be particularly useful as you suggest concrete steps each child can take to make his or her writing better. That is, if a writer's informational text is level 2, you and that writer can look at the descriptors of, say, structure for level 2 and note whether the writing adheres to those. If so, tell that child—or your whole class—if this is broadly applicable, "You used to structure your piece by . . . ," and read the descriptors from the prior level, "but now you are . . . ," and read the level 2 descriptor. "Here's a tip for how to make your writing even better. You could . . . ," and read from the level 3 descriptor. You might add, "Let me show you an example," and then cite a section of the level 3 exemplar text. Over the course of the unit, students will have the opportunity to keep tabs on their progress, checking their work not only against their original writing from the pre-assessment (which they can tape to the inside front cover of their notebook), but also against the checklist. They can notice both how they've grown as informational writers and what they can do to get even stronger.

At the end of the unit, you'll give the performance assessment again, using the same prompt to invite students to write another on-demand information text. You will be able to lay the two pieces alongside each other and mark them up, annotating in ways that show off students' new skills.

GETTING READY

As you prepare for the unit, it is important to gather materials that support your students both as scientists and as writers. First, you'll want to have books that support writers in these two roles. We suggest you comb your library, looking for a variety of books that support the content students will study over the course of the unit—that is, texts on forces and motion that describe push or pull, friction, and gravity. You will also want to have a variety of texts that will support your children as writers. Mentor texts to support scientific writing, in particular, are a plus. We suggest *Forces and Motion (Hands-On Science)* by John Graham and John Le Jars, which is in our text set. You will also want to choose an information book for Bend III of the unit, to use as a mentor text.

In addition to mentor texts, you will want to gather a variety of physical objects related to the study of forces and motion. Toy cars, ramps, meter or yard sticks, plastic spoons, cotton balls, masking tape, and rubber bands will support the work of the first two bends of the unit. Don't worry if you don't have all of these objects handy. Chances are, many of your students will be more than happy to add to the collection!

Finally, it is important to imagine the writing your students will be doing and to create some exemplar and demonstration writing pieces to use during your minilessons, conferences, and small groups. You may even want to create these in the same kinds of booklets your students will use to write their own books, so that these feel to kids like true models of what they might themselves write. Certainly, you will want to create a "bare bones" writing piece that could be used several different ways for revision. In this case, we recommend that you copy this piece so that you can use it in a variety of ways. Your writing is a key element for supporting writers throughout the unit.

Writing as Scientists Do

BEND I

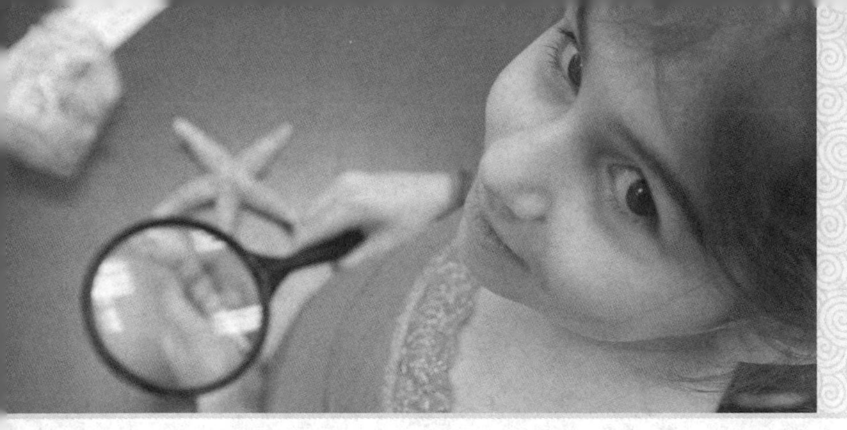

Session 1

Learning to Write about Science

IN THIS SESSION, you'll teach students that scientists study the world around them, pose questions and hypotheses, conduct experiments, and write about their results in lab reports.

GETTING READY

- ✔ Writing tools and clipboards
- ✔ A stack of four- or five-page stapled blank booklets with a picture box and six to twelve lines, one for each student (see Connection) and one for yourself (see Teaching and Active Engagement)
- ✔ Materials for the whole-class experiment, including a ramp, carpet, several meter sticks or yardsticks, and a toy car
- ✔ Chart with the scientific process (question, hypothesis, procedure, results, and conclusion)
- ✔ "To Write Like a Scientist" chart (see Teaching and Active Engagement, and Share)
- ✔ Post-it® notes for students (see Share)

Common Core State Standards: W.2.2, W.2.7, W.3.7, RI.2.1, RI.3.1, SL.2.1, SL.2.2, SL.2.4, SL.3.4, L.2.1.e, L.2.2, L.2.3, L.2.5.b, L.2.6, L.3.6

ONE OF THE WONDERFUL THINGS about writing is that it extends and amplifies any kind of learning. If you are a chef, writing can give your fans access to your recipes and your advice. If you are a dancer, writing can allow others to understand your art. If you are an economist, writing enables you to see patterns in data and to inform others of what you see. Information writing is an especially portable skill, one that can go with you into all the parts of your life, and that can be passed among all the members of your community.

Today opens a new unit, one that illuminates the processes and nature of information writing by inviting young people to write like scientists. Whenever writing is embedded in a discipline, the goal must first be for the writing to amplify the learners' understanding of that field, and surely today's writing, and all the writing that children do in this unit, will boost their knowledge of scientific processes in general and of the specific topic we've selected to study. We've chosen to focus on forces and motion within this writing workshop because this is a foundational unit in physical science that tends to be undertaught in today's elementary schools, and because physical science has special power in today's world. Then, too, it is easier for children to participate in the scientific processes when they are working with ramps, balls, and small machines rather than when they are working with animals and plants, because experimentation is easier, the results more forthcoming.

Although this unit is embedded in a writing workshop and we are certainly not suggesting that it take the place of other science work, today's session contains the three components that are part of most good science classes:

Students collect and look at data.

Students study and think about patterns in the data.

Students develop models or explanations of the data.

That is, students will not just be writing *about* scientific topics, they will be writing as scientists, using writing as part and parcel of their involvement in the scientific method. The intention is not just to teach concepts and processes of science, it is also to teach children how writing can extend their learning and allow them to develop—and then communicate—expertise. This unit focuses more on the writing than on the science. If studying force and motion isn't a good fit for you, you can transfer this teaching to another area of science. Specifically, during the first two bends of the unit, you'll teach children that they can write to hypothesize, to record their procedures, and above all, they'll write to grow theories about what they learn.

"Students will not just be writing about *scientific topics, they will be writing as scientists."*

Today's session is an unusual one. Fourth- and fifth-grade teachers refer to sessions like this one as "boot camp" because kids will go through a rigorous, intensive experience that functions a bit like a crash course. You'll abandon the traditional structures of a minilesson—although there is a teaching point and a merged sort of teaching/active engagement, you won't teach kids a three-step strategy that you demonstrate and then send them off to do independently. Instead, you will put them through a more complicated sequence that starts with asking and recording a question, then designing and conducting multiple trials of a simple experiment. Children will jot and sketch as they go, getting a four-page lab report booklet started during the meeting area, with their hypotheses on one page, their procedures on another, their results on a third, and their conclusions on a fourth.

Expect that most of this writing will be a very rough approximation of what your children will learn to write during the unit. For example, they will probably write procedures without any recognition that this is how-to writing that should be done in steps. Don't correct their ways, for now, as the upcoming sessions lead you to do so with some depth. For now, the goal is to quickly show children all the parts of the scientific method and the corresponding writing it yields. Throughout the unit, you may indeed find that students gesture toward the kind of scientific writing they will grow into later. Instead of being frustrated by their attempts, celebrate them, much as you celebrate brand-new writers' invented spellings, knowing that these pave the way for more sophisticated work down the road.

As on any opening day of a unit, you will rally kids to embrace the big work of this unit. You want your youngsters to be thrilled at the prospect of conducting science experiments and writing to figure out theories about how things move. You'll want them to dive in, approximating the work that scientists do when they write lab reports that contain hypotheses, procedures, and results.

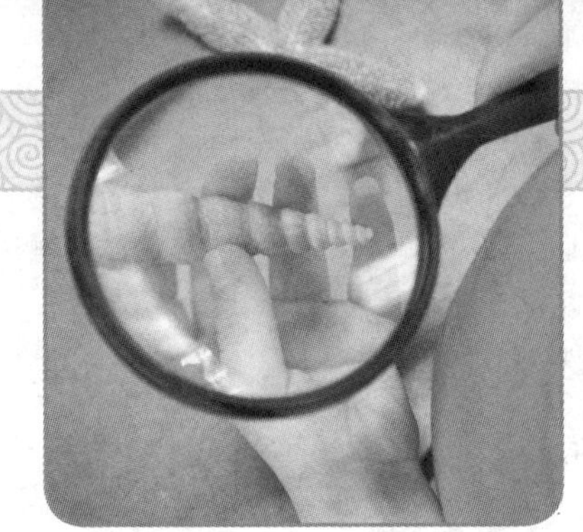

MINILESSON

Learning to Write about Science

CONNECTION

Ask students to visualize the kinds of writing work scientists do, and then describe that work.

As students arrived in the meeting area, I gave each one a blank lab report booklet, asking them to sit on it until the time when we needed it. "Today is an exciting day! You are about to learn a brand-new kind of writing. You learned information writing before, when you wrote about topics on which you were experts, but starting today, you will learn a whole new kind of information writing.

"In this unit you will learn to write in a way that scientists write. Let's think about what that means for a moment. Take the person who invented jelly beans—what kind of writing do you think she needed to do? And the person who discovered how to kill bacteria that makes people sick—what kind of writing did he do? And the team of scientists who are working on finding new forms of energy that don't destroy our environment—what kind of writing do they do? Right now, in your mind, picture a team of scientists at work. Thumbs up when you have an image in your mind."

I paused for a moment as I did this thinking myself. "Now tell your partner what you imagine those scientists doing and tell your partner what kind of writing they need to do."

I listened in as they talked briefly.

Confirm that scientists do write to plan, to record what happened, adding that they also write to teach. Explain that in this unit, children will first write like scientists do when the goal is to learn.

"Writers, I heard so many of you saying that you pictured people surrounded by special equipment! Sam said he pictured a guy with wild hair like Einstein, writing down a plan for designing a robot. Jamal said he pictured a scientist pouring foaming gel into her glass beaker and writing down what happened. You are right that scientists write down their plans, like Sam's imagined Einstein, and they write down what happens in their beaker full of foaming gel. And the important thing about this is not just that they write plans or write what happened—the important thing is *they are writing to learn more about things in the world.*

"Scientists do also spend a whole lot of time writing to teach other people what they've learned, and you'll do that work later in this unit."

◆ COACHING

I am always mindful that the ending of an old unit and the start of a new one allows children who haven't felt successful before now to have a new start. I emphasize the novelty, the freshness, of the new work to especially reinvigorate those children. And you will find that this unit does *appeal to youngsters who didn't take to the previous work on storytelling. Howard Gardner suggests that all human beings fall into one of two categories—the dramatists and the patterners. Give kids a pile of toy plates, and the dramatist will create a dinner party, complete with guests. The patterner will sort and categorize plates: big and small, chipped and flawless. This unit will appeal especially to the patterners.*

Establish the format of today's lesson: children will be guided through the process of conducting an experiment and writing within each step of that experiment.

"But we are going to start by writing like scientists do when they are writing to learn more about the world. And one way scientists collect information is to create experiments—to try things out and see how they go. Today we are going to do that in this workshop time. So, instead of a short minilesson and then a long time when you go back to your spots to write, today we are going to work in the meeting area for a long time, conducting an experiment together, and I will help you learn how scientists write as they go through all the parts of an experiment.

"By the end of the day, you will have written your first lab report, and, more important, you will have a sense of the way lab reports go so that you will be able to write these to help you learn from your own experiments."

Name the teaching point.

"Today I want to teach you that when scientists conduct experiments to learn about the world, they have a certain way they usually write—they use a lab report format. They record what they expect to happen in an experiment, and they record what they actually do in the experiment, then they record how things go and what they learn."

I can understand a lab format

TEACHING AND ACTIVE ENGAGEMENT

Teach through guided practice: Take children through the process of doing an experiment and writing a lab report. Coach them as they form and record a hypothesis, then conduct and record the experiment.

"Scientists often start by posing questions they have about the world around them. We are going to work on my question today so that you can see how this goes. Another day, you will start with your own question."

"Here's my question. While I write it down for myself, will you get a start writing it as well, on your first page of your lab report? That's where the question always goes." I turned and wrote, voicing as I did:

> Will this little car go farther off the ramp on carpet or on bare floor?

While the children continued copying the question, I said, "That is the question we need to figure out scientifically. I'm asking whether the differences in the surface—the bare floor versus the carpet—makes a difference in how far the car goes, and if so, exactly how much difference.

"Eyes up here, even if you are not done recording the question," I said. "The next thing a scientist needs to do after arriving at a question is to think and write what the answer *might* be. Will you do that with me right now? On the next page in your lab report booklet, will you hypothesize whether this little car will go farther on the carpeted floor or farther on the bare floor?

This is a complicated lesson with many parts to it. You will want to guard against letting yourself elaborate on this or that point, or gathering too many examples, or talking about what you are teaching. If the lesson ceases to be trim and spare and becomes more of a chain-of-thought creation, made with off-the-cuff associations, it could easily become a swamp. For this one minilesson, you might stick more to the plan than usual so as to keep up the pace and protect the cohesion of the plan.

try the counter by the window. instead of the bare floor

"What do you think the answer will be, *and why do you think it will turn out that way*? When you write a hypothesis you are writing the answer you expect to see based on all that you know." I added, "You may want to begin writing your hypothesis like this," and I jotted on the chart paper: "I think" and "My hypothesis is." "For now, will you jot your hypothesis about whether the car will go farther on one surface than another onto page 1 of the blank lab report book I gave you?"

As children jotted, I wrote:

To Write Like a Scientist . . .

1. Ask a question about how the world works.
2. Record a hypothesis, a guess.
3. How will you test it? Record your procedure.

"Turn and tell your partner your hypothesis," I said, and then listened in as children talked. After a minute, I asked for writers' attention. "Many of you think that the car will go a different distance on the carpet than on the floor—and usually you think it will go less far on the rug. Some of you even mentioned that the bumpiness of the carpet will make the car stop sooner. Each of these predictions is a hypothesis."

You will note that I haven't given the kids much help with any of this. The chance that their work will resemble the work of most scientists' lab reports is not high! The intent of this lesson is to show the flow of an experiment and the parts of a lab report, not to teach all about each of the parts. That instruction will begin on Day Two, with an invitation to kids to revise. It is fine, then, that these early drafts are less than ideal—that will heighten the need for revision!

Channel children to plan and record a procedure for testing their hypothesis.

"The next thing that scientists tend to do after identifying a problem and hypothesizing is ask, 'How can we test this?' Scientists think about the procedure they will follow to test out the question—and that's hard.

"In a minute, I'm going to ask you for help conducting this experiment," I said. "Right now we have to figure out what the experiment will be. How will we test out whether the car will go farther on the wooden floor or farther on the carpeted floor? Start designing the experiment out loud, with your partner." I let children talk for a moment.

Instead of gleaning suggestions from the whole class, I reconfigured the group. "Boys and girls, can you position yourself on the perimeter of the carpet?" As the children moved so they were framing the meeting area, I asked for student volunteers to help set up the experiment. Speaking to the rest of the class, I said, "Scientists always plan out what they will do in an experiment, so on page 3 of your lab report, will you use story boxes to quickly draw and label what you think we should do with all this stuff?"—and I gestured to a ramp, a car, the meter sticks and the like. Then I said, "Meanwhile, I need the four volunteers up here to share ideas for the procedure we should follow."

When choosing volunteers, we strongly suggest you recruit children who haven't yet found that writing is their "thing." Writing has a great elasticity, and often people who do not care for literary writing will readily take to informational writing. If this unit recruits youngsters who hadn't previously seen themselves as writers to identify with this kind of writing, that is a big deal.

The four volunteers sprang to the front of the rug area and conferred while others wrote. Soon they'd put the ramp at one end of the meeting area, on an incline, and laid out the meter sticks on the carpet, ready to measure the distance the car traveled once it left the ramp.

Ask the volunteers to share their planned procedures, naming the precise steps they will follow and to then conduct one leg of the experiment in front of the class. Channel children to record the results, including the unit of measurement.

"So, class, let's listen to these four scientists' planned procedures and see if the rest of us agree with their plans." Turning to the foursome, I asked, "What will your procedures be?" One answered that first they'd have the car travel on the carpet, then on the bare floor.

"So what is the first step?" I asked, and we learned that first a child would stand, hand over the top of the ramp, and release the car from that spot at the top of the ramp—not giving it a push. Then another child would measure the distance traveled. "Scientists, I appreciate the precise directions. That's very scientific," I said. "When the procedures are precise like this, when other people follow them, they should get the same results." Then I told the class, "If your storyboard pictures of the procedures aren't equally precise, add details to your drawings to remind you of the details. Later you can write this page up."

The four children at the front of the meeting area meanwhile did the experiment, with one rushing to the meter stick to call out, "Fifty-seven."

"Fifty-seven what?" I replied, leading the child to clarify that the car had traveled fifty-seven centimeters. "Scientists always include the unit of the measurement. Fifty-seven miles would be very different than fifty-seven centimeters!" Then I reminded children that on the next page, they needed to record their results. "For now, quickly draw and label these. You will want to leave space to write in the results you got when the car was on the carpet and another space to record results you got when the car was on the bare floor."

At this time children will be rehearsing the language that they will later record. We know that it will take some second-graders a bit longer to write each step in the procedure. We want them to quickly collect and remember the important parts of the procedure to later record in detail during independent writing.

Channel the class to conduct multiple trials.

"Scientists try their experiments several times to make sure they get consistent results. So as I write that last step, we'll try your same experiment two more times, with everyone recording the results each time."

4. Conduct multiple trials, and record your results.

Different children released the car the next time, and the next. The car traveled fifty-seven centimeters on the first trial, fifty-four centimeters on the second trial, and fifty-three centimeters on the third trial. The class briefly discussed that these results were similar and wondered why they were not exactly the same. The children recorded those results in quick sketches and notes on their results page.

Debrief—reiterate for the class what the volunteers did that you are hoping all writers have learned to do.

"Writers, did you see that when scientists want to learn by experimenting, they go through a process that involves some thinking and talking and doing, then some writing. When writing about an experiment, your writing is expected to have different parts." I pointed to our chart. "A question, a hypothesis, a procedure, some results, and a conclusion."

LINK

Set children up to conduct and record the second leg of the experiment with more independence, while still in the meeting area, contrasting the results from this trial with those from the earlier trial.

"So, writers, scientists, you can't tell if the car is going farther on the carpet or on the bare floor until you have collected data from the bare floor as well as the carpeted floor. So, the next step of our procedure will be 'Repeat steps above, this time with bare floor.'" I recruited four new volunteers and said to them, "We need to test on a bare floor, so get started conducting the experiment!" The rest of the class looked up to watch, and with great suspense, one child released the car at the top of the wooden ramp once more, and other students measured the distance traveled, and everyone recorded the data. Again this was repeated. Again children recorded results.

"Writers, today's meeting was extra long, and so you don't have a lot of time to turn your sketches, labels, and important numbers into a lab report," I said. "But in fifteen minutes, I'm pretty sure you can get every page written out, with all your sketches and numbers turned into sentences. I would start by rereading the first page and then remember the experiment and write, write, write. Thumbs up if you are ready to write your first lab report in your life." Once I saw thumbs going up, I said, "Okay, get going!"

When you shift from the demonstration to debriefing, students should feel the different moves you are making just by the way your intonation and posture change. After most demonstrations, there will be a time for you to debrief, and that's a time when you are no longer acting like a writer. You are the teacher who has been watching the demonstration and now turns to talk, eye to eye with kids, asking if they noticed this or that during the previous portion of the minilesson.

Teachers, if you doubt that children will be able to conduct this experiment and write this much in a single writing workshop, know that time and again we find children rise to the occasion. If you need to extend your writing workshop by five minutes to feel confident about this, do. The important thing is that you convey in your voice total confidence that kids can pull this off. As they write, use voiceovers to keep their hands flying. "You should be on your second page by now." "Write like the wind. Don't stop to reread. Just rush your ideas onto the page." Teachers who were very skeptical have come away saying, "I wouldn't have guessed they could do so much." The secret is to convey utter conviction that kids can conduct an entire experiment and write a five-page booklet version of a lab report in the allotted time. (Convey this even if you are faking that confidence!)

Supporting Engagement

ALWAYS, WHEN YOU BEGIN A NEW UNIT, circulate quickly to make sure your children understand the gist of what is expected of them well enough that they can roll up their sleeves and get involved. Always, your goal will be to issue a wide invitation into this new work, not worrying too much whether every child's work matches your expectations. There is time enough later to ratchet up the level of the work. For now, you mostly want to be sure that every child can carry on with independence; then you are free to teach. But if many children aren't clear what to do, and they are swarming you to ask which page contains what kind of writing, or what your expectations are for this or that aspect of this lab report, then you won't have spare energy or time to observe and to teach, based on what you see.

Be clear to yourself, then, what the writing is that you expect children to do while with you in the meeting area and what the writing is that you expect children will do once you send them off to work on their own. Usually when we have taught this session, some children get a fair amount of writing done in their lab report booklets while they are sitting on the rug and the class is proceeding through the experiment. But more children tend to write the question, the hypothesis, and then some labeled drawings or notes only for the procedures. Either way, you are presumably imagining that once children leave the meeting area, they'll reread their entire booklet and make sure that they have written everything they want to say on each page. Some will rely on labeled drawings and a few sentences to convey their meaning, and others will write much more.

Once you've channeled all your children toward capturing the science experiment in their lab report books, you'll be able to lift the level of their work. One of the most urgent and important goals is to support children in writing longer and more. You can use voiceovers to help with fluency. "Your pen should be flying." "Say a sentence, then write that sentence before stopping." You'll also want to encourage children to use the science terminology they are learning starting with terms such as *conclusions* and *procedures*. Be sure to celebrate children who are brave spellers, tackling words they haven't entirely mastered.

But meanwhile, you'll also want to lift the level of students' engagement with science. The most important way to do that is to pay attention yourself to the content and show in every way possible that you find friction and gravity and motion fascinating. If children are acting silly, as if the invitation to engage with science is an invitation to turn school into playtime, act astonished, and turn the focus immediately toward the intellectual work.

MID-WORKSHOP TEACHING

Drafting Results and Conclusions

"Scientists, eyes up here." I waited until the room was silent. "Some of you were probably wondering how to fill in the last page of your booklet. After scientists finish conducting experiments and collecting data their job is not done. Scientists write about their results, what happened, and then about making sense of it—the conclusions. In this case, when you are writing the results you are going to write two sections of results—one section will be what happened when the car was on carpet. The second section will be what happened when the car was on the bare floor.

"Then, you will write the final page of the lab report—the conclusion. In this section, the scientists answer the question asked at the start and describe what they learned. But conclusions aren't just the results—they aren't just the numbers. The conclusions are also the thoughts that scientists have about why the experiment turned out as it did. Scientists don't just write what happened, they also write about what they think about what happened. They write what surprised them and what they make of those surprises—and they write about questions this experiment raises."

As Students Continue Working...

"Writers, remember to write about everything that we did in the experiment! You don't want to leave out important information!"

"I just noticed that Jesse opened up his booklet and went back up to the writing center to get more pages. He didn't just add them to the back of the booklet. He needed more pages in the middle of booklet and added them there. Remember, if you need more space to write, just help yourself and keep writing!"

SHARE

Writing Like Scientists

Ask writers to show partners where they did each of the kinds of writing you have explained and listed as the components of a lab report.

"Writers, come quickly with your writing, a pencil, and a Post-it note because we have a lot to talk about, and as soon as you get here, you'll show your partner where you did each of these things. In fact, will you write a number 1 on the page where you did #1, and write a number 2 on the page where you did #2, and so forth.

To Write Like a Scientist . . .

1. Ask a question about how the world works.
2. Record a hypothesis, a guess.
3. How will you test it? Record your procedure.
4. Conduct multiple trials, and record your results.
5. Analyze your results, and write a conclusion.

After giving children a minute to do that work, I said, "So here is my question. Of these parts of your writing, which is the strongest? Which part are you most proud of?" I gave children some time to think about that. "When you find that part, put your Post-it note next to that spot.

"Writers, share the part of your writing you are most proud of. After you read the part aloud to your partner, make sure you tell your partner *why* this part is your best. What made the part you selected into a great piece of writing?"

I listened in to a partnership discussing their writing. Michael shared, "I worked really hard here." Michael pointed to his hypothesis and said, "I said my idea and then I said the 'because.' I explained why I guessed that the car on the bare floor would go farther."

Then, I directed my attention to another partnership. "I did math here," Rebecca said as she pointed to her calculation. "I knew that the distance the car moved on the floor was much longer. I wanted to know how much longer. I subtracted the two numbers," said Rebecca as she proudly pointed to her calculation.

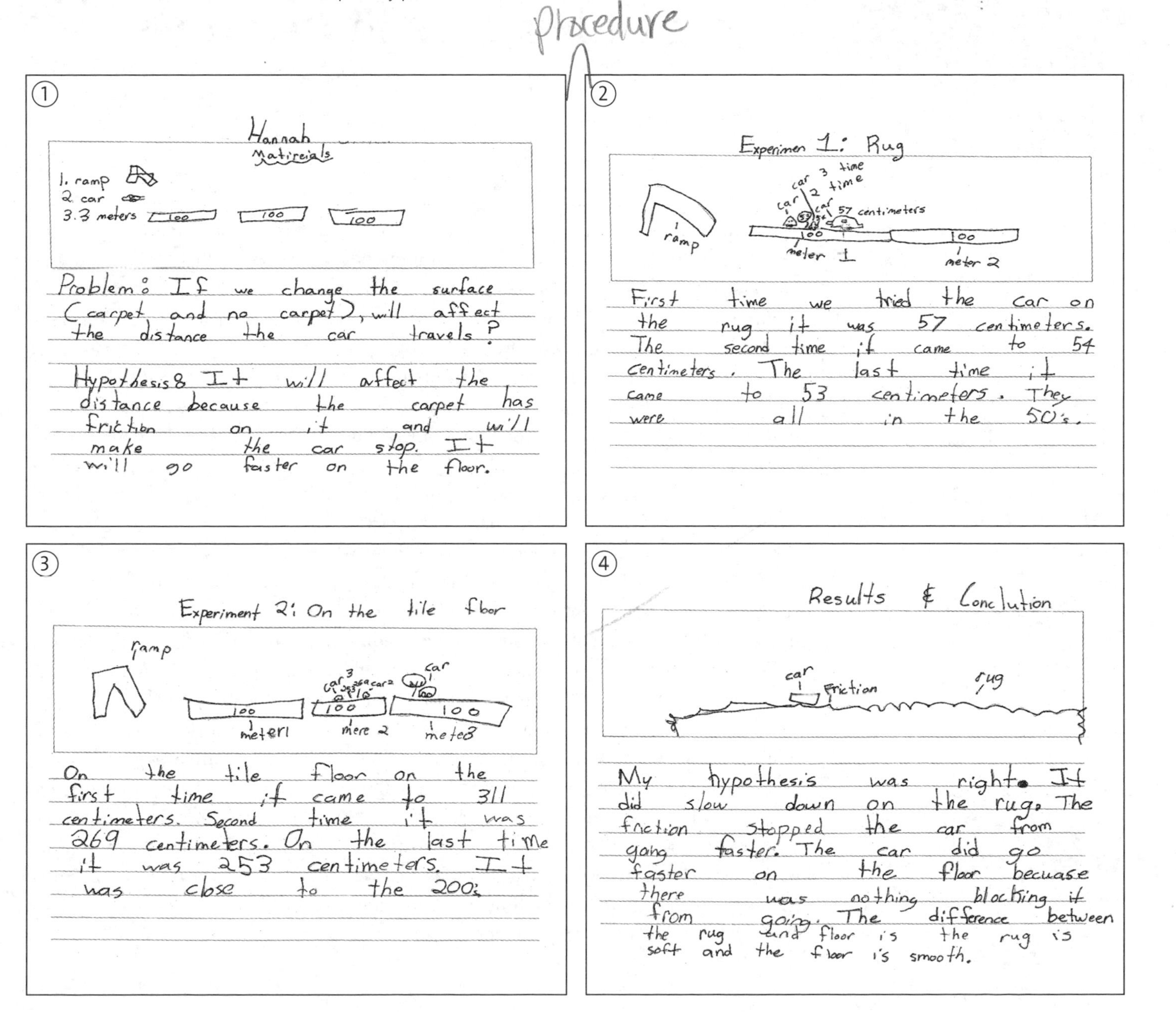

Procedure

① Hannah

Matireials

1. ramp
2. car
3. 3 meters 100 100 100

Problem: If we change the surface (carpet and no carpet), will affect the distance the car travels?

Hypothesis: It will affect the distance because the carpet has friction on it and will make the car stop. It will go faster on the floor.

② Experimen 1: Rug

ramp; car 3 time; car 2 time; car; 57 centimeters; 100; meter 1; 100; meter 2

First time we tried the car on the rug it was 57 centimeters. The second time it came to 54 centimeters. The last time it came to 53 centimeters. They were all in the 50's.

③ Experiment 2: On the tile floor

ramp; car 3; car 2; car; 100; meter1; 100; mere 2; 100; mete3

On the tile floor on the first time it came to 311 centimeters. Second time it was 269 centimeters. On the last time it was 253 centimeters. It was close to the 200's.

④ Results & Conclution

car; Friction; rug

My hypothesis was right. It did slow down on the rug. The fraction stopped the car from going faster. The car did go faster on the floor becuase there was nothing blocking it from going. The difference between the rug and floor is the rug is soft and the floor is smooth.

FIG. 1–1 Hannah's lab report

Session 2

Studying a Mentor Text

Procedural Writing

IN THIS SESSION, you'll teach students that writers study mentor texts when learning to write a new kind of writing, like procedural writing, asking what the author has done that they could try as well.

GETTING READY

- Students' booklets of experiments they drafted in the previous session (see Connection)
- Materials used in the experiment from the previous session: ramp, carpet, measuring sticks, and the toy car, displayed in the meeting area
- "In Procedures" chart (see Teaching and Active Engagement, and Mid-Workshop Teaching)
- Mentor text that has a procedural page; we use *Forces and Motion (Hands-on Science)* by John Graham (2013), "Floating and Sinking Experiment" (see Teaching and Active Engagement)
- Blank pages to add to booklets placed on writing tables—one for each student (see Link)
- Enlarged copy of the Information Writing Checklist, Grades 2 and 3 (see Share)
- Small copies of the Information Writing Checklist, Grades 2 and 3, one for each student (see Share)

Common Core State Standards: W.2.2, W.2.7, W.2.8, RI.2.1, RI.2.6, RI.2.7, RI.3.1, SL.2.1, SL.2.2, SL.2.4, SL.3.4, L.2.1, L.2.2, L.2.3, L.2.5.b, L.2.6

LAST YEAR, WHEN CHILDREN WROTE nonfiction chapter books, they learned that some pages of their books were apt to be structured like lists. The glossary, for example, is usually a list, and pages in which an author describes different examples or different parts might also be organized as lists. Still other chapters in the children's first-grade teaching books were actually how-to writing. A chapter in a dog book that was called "How to Get a Dog" may have been part of an all-about teaching book, but that page, in and of itself, would have been more accurately described as a how-to text. Today's session harkens back to that earlier lesson, reminding children that when they write information writing, sometimes the different parts of that writing require different structures. And specifically, lab reports contain a section (for these children, a page) called "procedures." Might it be possible to reread that page and to ask, "What is the structure, the genre, and the kind of writing that an author needs to be doing when writing this part of a lab report?"

This session is designed as an inquiry minilesson in which you invite youngsters to study and emulate a mentor text. Specifically, students will study and name what the mentor author has done in the published procedure page and they will ask that familiar question, "What has this writer done that I, too, could try?" Then they will rewrite any page (or pages) on which they'd written about their procedure on the previous day.

Embedded in this is an expectation that writers revise in large-scale ways, pulling out one page of a book and substituting it with another. You'll put your full weight behind the children making large-scale revisions and won't allow for as much choice as usual, because your unit will be new enough that children don't yet have a repertoire of optional ways to proceed. The result is that the revisions will be a bit lockstep for now, with the whole class doing similar work each day, but soon your children will be progressing with more independence. If there is more time, you can, of course, encourage children to return to and extend any part of their lab report.

MINILESSON

Studying a Mentor Text

Procedural Writing

CONNECTION

Help children understand the purpose of writing up their experiments with exact, precise information. Then name the question that will guide the inquiry: What does a scientist do when writing the procedure section of his or her lab report? How do procedures go?

With the materials from yesterday's experiments set up on display, I called children to join me at the rug with their writing folders and booklets in tow. Gesturing toward the display, I began, "You know, writers, I have to tell you that just as you were curious when you first saw these cars, ramps, and meter sticks yesterday, so were the many visitors who have been coming in and out of our classroom. Mr. Feeney came in to borrow a book and asked me, 'What are you doing with those toys?' and when I told him about our science experiments, he wondered if his class could try them.

"I realized that these people (and others) might actually be interested in trying out some of our experiments for themselves, which could allow us to get more data, to conduct multiple trials. We could end up making a big chart of the results people across the school get when they try this experiment. Perhaps if Mr. Feeney's class did this with a different vehicle, their results might be a bit different. Or they might be the same.

"The important thing will be for you to share your procedures with Mr. Feeney's class and with others in the school who want to try the same experiment. Before you do that, however, you'll want to be sure that your procedures are written as scientists would write procedures, so that when the different classes try the experiment, they are all following the same procedures. What if Mr. Feeney's kids started their car like this?" and I whipped an invisible car off from the starting line, suggesting an altogether different start to the trial.

"Today, instead of a regular minilesson, let's figure something out together! You and I both know that there is a way to learn what scientists do to write their procedures—right? How can we figure this out?" The children's hands shot up and they announced that they could study how published scientists write their procedure sections. I agreed.

Name the question that will guide the inquiry.

"The question we will be researching today, then, is this: What does a scientist do when writing the procedure section of his or her lab report? How do procedures go?"

◆ COACHING

It is impossible to overemphasize the value of creating a purpose, a real audience, for children's writing. You will see that throughout this unit, we try to embed the writing that students do into the real world, and today is no exception. We encourage you to recruit a colleague to do as Mr. Feeney has done. If all the second grades across the school are doing this work, for example, we hope you can stagger the schedule a bit so that one teacher can learn from another. And the truth is that the reason that procedure pages need to be extremely detailed and precise is for exactly the reason that you spotlight today.

You'll learn from comparing and contrasting this lesson with inquiry lessons in other books. Doing so will help you realize how easy it can be to write minilessons for yourself. Each follows a very similar template.

It is helpful to have recorded the question in advance and to reveal it as you read it aloud.

TEACHING AND ACTIVE ENGAGEMENT

Set children up for a mini-inquiry, preparing them to study a mentor text for something they could try in their own writing.

"Writers, when I need to learn to write in a new way, I study that new kind of writing, thinking about how it is made and what the parts are of that writing. You can do this, too! I thought we could study this procedural text from *Forces and Motion.*" I displayed the text, then said, "As I read the text, will you ask yourself, 'How does a procedure go?' 'What did this writer do that I could try too?'"

Floating and Sinking

YOU WILL NEED

- *A Two-Liter Plastic Bottle (with cap)*
- *A Flexible Straw*
- *A Large Paper Clip*
- *An Aluminum Pie Plate*
- *Modeling Clay*
- *Scissors*
- *A Bowl of Water*

1. *Cut out the shape of your diver from the pie plate. Make him tall and thin, about 2½ in. (7 cm) by ¾ in. (2 cm), so he will fit through the neck of the bottle.*
2. *Bend the straw at the neck, then cut it so you have a U-shaped piece about 1 in. (2.5 cm) long. Slide the open ends of the straw onto the two ends of the paper clip.*
3. *Gently slide the paper clip and straw between the diver's legs and up onto its body. The straw should be on his back, bent at the top behind his head.*
4. *Make diving boots out of modeling clay and put them on his feet.*
5. *Try floating your diver in a bowl of water. Carefully adjust the amount of clay on his boots until he just floats.*
6. *Fill the bottle with water and put the diver inside. Make sure the bottle is full to the top, then screw the cap on tight. The diver should float to the top.*
7. *Squeeze the bottle. The diver will sink to the bottom. Let go, and he'll float up. With care, you can make him float as deep as you like!*

If you do not have a document camera or another way to display an enlargement of the page, perhaps you can purchase a few copies of the book, giving one copy out to every cluster of children. Each writer need not own his or her own copy, but the children do need to see the layout of this page, especially. It is important to know that you need not use this exact text! Any short example of a well-written procedure, on this topic or another one, will do. A well-written recipe, for example, could work here.

Introduce the mentor text, encouraging children to study it.

"So writers, before you read this experiment from *Forces and Motion* would you study the whole page? Think about all the things that the author has done that you could try in your procedure page. Put on your thinking caps." I left a long moment of silence, and during that time, I studied the page.

"Turn and tell your partner some things that you are noticing.

"Now that you noticed a few things, will you and your partner actually read the words and see if you can notice *more* things that the author has done that you could do as well?"

For a minute, I did the same work as students were doing, then I began listening in and coaching their talk. As children worked, I voiced over: "I like how Jenna and Mark are going back to reread the first section. They are reading each part, then stopping to ask, 'What is this writer doing in that part that I could try?' That's a strategy to use for learning a lot from one page of writing."

Chart children's observations about the mentor procedural text.

Reconvening the class, I asked. "Writers, what did you notice that you could do on the procedure page of your lab report?" On the easel, I wrote: "In Procedures . . ."

Jesse said, "There is a section that tells you what you need." I wrote, "Make a 'You will need' section" on the chart. After a few more children shared, our chart looked like this:

In Procedures . . .

- Make a "You will need" section.
- Draw pictures that teach with labels, details.
- Number the steps.

Expect students to notice only the most obvious elements of a procedural text. They'll probably point out the presence of numbered steps, for example. That is okay! You can coach children to look more deeply later. You might also have students who become distracted by the experiment in the book and who share observations about the content of the text rather than the writing techniques. In that case, you will want to remind them that they are studying how *the author/scientist is writing, not* what *he is writing about.*

LINK

Explain that writers will all begin anew, writing a whole new procedure page, and set them up to imagine how it will be much better.

"Writers, now the hard part. You ready? I'm going to give each one of you a *new* procedure page, a blank one, and you will have the chance to write that page all over, from the beginning, making it so so so much better. Right now, think about all the things you noticed in the book *Forces and Motion*. In your new procedure page, will you do some of what that grown-up author did? I know you are only seven, but still, I'm thinking you can write like pros. In your mind, plan what you will do differently."

Reiterate the importance of precise procedures, and channel all children to disassemble their books, removing their old procedure pages and replacing them with blank pages.

"Writers, remember that other kids in this school want to replicate your experiments. If you don't have very clear procedures, kids in room 306 might end up saying that their car went clear across the classroom, and we might think there is something different about room 306, when really, perhaps we find out later that they didn't use a car like ours, instead they used a car with a battery on it.

"Right now, while sitting here in the meeting area, why don't you take apart your book (removing staples) and take away your old procedure page. After you have done that, go to your work spot and find a few different versions of a new, blank procedure page that I put in the middle of each table. Although we are focusing on making our procedure pages clear, remember that you can go back to other pages and revise them, too, to make the writing on those pages even more clear and precise."

You always want to be ready to improvise. So if you don't think your kids can take out the staples from their books, ask them to fold up their old procedure page, or cross it out, or label it Draft 1 and label the new version of the page, Draft 2. Invent!

Channeling Students to Use Mentors from Start to Finish

On any day when your instruction channels youngsters to notice the mentor text, it helps to carry a few copies of the text with you as well as Post-it notes to leave with small groups.

Because you asked children to take apart their books while sitting in the meeting area and to remove their old procedure page, directing every child to draft an entirely new page, you'll probably find that you don't need to spend most of your energy cajoling kids to do large-scale revisions. Instead, you can help them to write this new page as well as possible.

When your aim is to teach in ways that lift the level of student work, it helps to teach youngsters to make one plan, then rethink that plan, then make an improved plan. Children tend to latch onto one plan and to then put that one plan into operation. Sometimes the result is an improvement on the original draft, but sometimes it's not. On the other hand, when you channel children to draft and share and discuss possible plans, postponing action on those plans to allow for at least a few minutes of deliberation gives you time to hear and coach into children's plans. Today, then, you'll probably want to help several small groups to share and strengthen their planned revisions.

How will you help them improve their plans? Certainly you can make sure they include all parts of the page, that they use domain-specific vocabulary words, that they make their pictures instructional, that they provide sequenced and specific information, and that they add tips.

By the time one group of children has shared and improved upon their plans, most children will be well on their way—and you can support another small group working on plans. If you find that most children have started writing their drafts, you'll want to recruit a cluster of them to begin revising those drafts even before they finish their work. Collect those children—perhaps signaling, "Come with me," conveying that time is of the essence—and then remind them that writers know it is helpful to shift

MID-WORKSHOP TEACHING

Noticing More in the Mentor Text

"Writers, can I stop you?" I continued, "Are you ready for another challenge, an even harder one? You know that word—*revision*. Well, the *re* in *re-vision* means 'to do it again': think of the words *re-peat*, *re-heat*, *re-play*. And *re-vision* means to vision again, or to see again. One of the ways to become a good writer is to learn to read and revise, to see again. You can see the mentor text again, and if you really look closely, you will see more.

"Let's try it. Let's all just reread step 1 of this text together, and this time, instead of seeing that it is written in steps—a pretty obvious thing to notice—will you see if there are some details about the *way* this is written that you notice?" I reread the text aloud.

1. Cut out the shape of a diver from the pie plate. Make it tall and thin, about 2½ in. (7 cm) by ¾ in. (2 cm), so it will fit through the neck of the bottle.

"What *exactly* has this author done that could teach us more?" Soon we'd added "Include detailed measurements" and "Tell not only what to do, but how to do it" to our "In Procedures" chart.

- Include detailed measurements (2½ in.).
- Tell not only what to do, but how to do it.

"If any of you finish your procedure page, look at the rest of your pages. If you come to some ideas for how to make them even better, do it! And, remember what you did because we'd love to learn from you."

often from being a writer to being a reader. "Pretend you are one of the kids from room 306," you might say, "and you want to do this experiment. Read it, and see if you can figure out how to do each step." You might remind them of how, when they were in kindergarten, they wrote the steps to making a peanut butter sandwich, only the steps were so vague—"Put the peanut butter on the bread"—that they ended up with a jar of peanut butter perched on a loaf of bread. Say, "Writing procedures is really another way of saying writing directions, and it is not easy to make your writing clear enough that readers know exactly what to do."

If you have time to lead more small groups, you can also teach writers to reread the mentor text or other books containing experiments, noticing yet more that they could try. This allows you to support children in naming strategies they admired when they studied the mentor text. For example, Peter studied a mentor text that contained several procedural pages, each containing lots of text features. When I asked what he'd been working on, he touched his own sheet of paper, and said, "I put a 'You will need' box, and then I started writing steps. Then I wanted to add warnings and tips, so you would know how to do the experiment the *right* way."

I gestured for him to continue talking.

"And, well, I'm running out of space on this page, so I can't fit in all the steps. But I really want to put the steps in boxes, like in the book," he answered.

I studied his work and said, "I can see that you want your procedure page to be just like the mentor text. You want to include *all* those special things you noticed in the book. I appreciate that you are closely studying what another writer has done almost like it is a To-Do list. You have developed a system that makes it so that you learn not just one thing but *many* things from your mentor. Given that you have figured this out, I bet you can also figure out how to get yourself more paper. Am I right?" Peter nodded, and started back to work.

My hypothesis is that the car will go faster on the bare Floor.

1. Put ramp on carpet. Release car down the ramp and record how far it goes
2. Put ramp on bare floor. Reales car and record how far it goes (reapet 3-4 times).

FIG. 2–1 Magali's procedure page

scientist Magali

ramp

car

yard stike

You will need for this experament a ramp, a car, and, a yard stike. 1st place ramp in the Middle of the carpet then you place a toy car on the ramp on top of the ramp and poot the yard stike at the end of the ramp, then realease the car, and reacord how fast it went.

2nd place the ramp on bare floor. Warning: Do not place ramp near carpets then poot the yard stike at the end of the ramp, then place toy car at the top

FIG. 2–2 Magali revises using the mentor text.

of the ramp and realease the car, and reacord how far it went.

3rd record wich time you realeased the car went faster. On the carpet or the bare floor?

FIG. 2–3 Magali finishes the procedure.

SHARE

Self-Assessment

Introduce students to the Information Writing Checklist for Grades 2 and 3.

As children settled in to the meeting area, I studied the Information Writing Checklist that I'd displayed, building excitement around it by attending to it myself. "Writers, I have this new checklist for second-grade information writers that I think is so cool because it also has third-grade goals on it! You can use it to check that you are doing everything you've already learned as second-grade information writers, and to set new goals for yourself this unit. Just like your Narrative Writing Checklist in the last unit, this checklist has new headings for the third-grade expectations." (The Information Writing Checklist, Grades 2 and 3 can be found on the CD-ROM.)

"This is what you are going to do: Check your writing to see if you are doing *everything* you've already learned, and if not, star those items as goals you want to work toward. Right now, will you start with reading the items under 'Structure' with your partner and then dig into your writing to see which of those items you have done and which ones will be new goals for you this unit?"

Information Writing Checklist

	Grade 2	NOT YET	STARTING TO	YES!	Grade 3	NOT YET	STARTING TO	YES!
	Structure				**Structure**			
Overall	I taught readers some important points about a subject.	☐	☐	☐	I taught readers information about a subject. I put in ideas, observations, and questions.	☐	☐	☐
Lead	I wrote a beginning in which I named a subject and tried to interest readers.	☐	☐	☐	I wrote a beginning in which I got readers ready to learn a lot of information about the subject.	☐	☐	☐
Transitions	I used words such as *and* and *also* to show I had more to say.	☐	☐	☐	I used words to show sequence such as *before, after, then,* and *later.* I also used words to show what didn't fit such as *however* and *but.*	☐	☐	☐
Ending	I wrote some sentences or a section at the end to wrap up my piece.	☐	☐	☐	I wrote an ending that drew conclusions, asked questions, or suggested ways readers might respond.	☐	☐	☐
Organization	My writing had different parts. Each part told different information about the topic.	☐	☐	☐	I grouped my information into parts. Each part was mostly about one thing that connected to my big topic.	☐	☐	☐
	Development				**Development**			
Elaboration	I used different kinds of information in my writing such as facts, definitions, details, steps, and tips.	☐	☐	☐	I wrote facts, definitions, details, and observations about my topic and explained some of them.	☐	☐	☐
Craft	I tried to include the words that showed I'm an expert on the topic.	☐	☐	☐	I chose expert words to teach readers a lot about the subject. I taught information in a way to interest readers. I may have used drawings, captions, or diagrams.	☐	☐	☐

"Writers, I am going to keep this checklist up to remind you of all the things you should be working toward in this unit. In a few days, we are going to return to this checklist so that you can keep track of how you are growing as information writers!"

Session 3

New Wonderings, New Experiments

IN THIS SESSION, you'll teach students that scientists—like writers—go through a process. And just like writers, scientists come up with their own ideas for what to write about. They decide on a question they want to find out about and then plan and test their question with an experiment, recording all the steps as they go.

GETTING READY

- Students' booklets of experiments from the previous session, a blank booklet, and a writing tool
- Your own lab report from the previous session to use during the demonstration (see Teaching)
- Materials like the ones used in the experiment from Session 1: ramp, carpet, measuring sticks, and toy car—one set for each partnership
- Various materials that students can use for their invented experiments (blocks, different-size toy cars, books, math manipulatives) (see Teaching)
- "To Write Like a Scientist" and "In Procedures" charts, displayed for student reference
- A pointer, baton, or long pen for one partnership to use to demonstrate how they conducted their experiment (see Share)

Common Core State Standards: W.2.2, W.2.5, W.2.7, W.2.8, W.3.4, RI.2.1, RI.3.1, SL.2.1, SL.3.1.b, L.2.1, L.2.2, L.2.3, L.2.6

As mentioned earlier, this unit has been designed to reflect informed principles not only of writing education but also of science education. In too many classrooms, science education mostly involves teachers reading aloud or paraphrasing expository books to pour some information about science-related topics into children's minds. If children are involved in any experiments at all, they follow the prescribed steps as if they were baking and following a recipe. If there is any thinking about the procedures in the experiment, it tends to be the teacher who does that part, lest children produce erroneous explanations and conjectures.

The science education that is featured in this unit comes from a different place. We do encourage teachers to give mini-lectures, to show videotapes, and to read aloud expository texts because content knowledge matters—and you'll see more of that soon within this unit. But we also think that just as children need to participate in the writing process, writing like professional writers write, so, too, they need to participate in the scientific method, working as professional scientists work. And that means that kids need invitations to design an experiment, to think about what the results are apt to be, to plan procedures, to conduct multiple trials, and to study patterns in the data they collect, theorizing about those data. What's more, we know that seven-year-olds can do all of this—in their seven-year-old way.

Granted, the experiment a child designs may not be what you and I had in mind. A child might ask whether the distance a car travels would be influenced if the car were missing one wheel, or if it matters that the vehicle that travels down the ramp might not be a car at all but a teacher's glasses. The kids might wonder whether whispering encouraging words accelerates the movement. The trials a seven-year-old imagines might not be anything close to what you or I would imagine, but the important thing is that a seven-year-old can proceed through the scientific method, testing a hypothesis and recording each step of his or her work. And over time, children can be taught that the experiments they imagine don't just come from the thought, "Wouldn't it be cool if . . ." but instead, they come from an emerging theory that explains the previous results. That is, if the car does not go as far on the carpet as on the bare floor, the next experiment might logically involve a thicker carpet. If the distance a car travels seems like it varies according to the vehicle's size, then it logically follows that there could be experiments sending larger and smaller objects down the ramp.

"Just as children need to participate in the writing process, they need to participate in the scientific method, working as professional scientists do."

This session emphasizes science more than writing, but of course the two become inseparable. In this session, children are invited to improvise a follow-up experiment involving a variation on the initial cars, planks, and measuring stick experiment. Some will probably send a shoe, not a car down the ramp, some will try for a larger or a smaller car, some will adjust the incline of the ramp. Whatever they do, your hope is they do it with energy and precision, and that they remember to record every step of the way—putting their hypotheses, their procedures, their results, and their conclusions into a second lab report.

Later, still, you will teach in ways that lift the level of all this work so we again encourage you to refrain from worrying if neither the writing nor the experimentation meet your expectations just yet.

MINILESSON

New Wonderings, New Experiments

CONNECTION

Tell children that just as they revised their lab reports, scientists also revise their experiments. Rally kids to design and conduct their own variations on the class's first experiment.

"Writers, yesterday you revised your first lab report. As you continue to learn more, you will discover more ways you can revise that first lab report. But today I want to tell you that just as you wrote your procedure section one way and then you looked back at it and thought, 'Wait, I have another idea for how this could go!' so, too, scientists do their experiments one way, then look back on what they have done and think, 'Wait, I know another idea for how this could go.'

"If you were grown-up scientists, you wouldn't *just* send a car down a ramp to see how far the car travels on a carpeted floor compared to how far it travels on a bare floor. Grown-up scientists notice patterns in the results, think about what those patterns might mean, and then after they get hunches, they change things around, trying that experiment over again but in different ways. They say, 'What if . . .'"

> What if the carpet at the end of the ramp had more texture, was thicker? Would the distance the car travels change?
>
> What if the car were heavier (or lighter)? Would the distance it travels change?

"But you all are too young to plan and conduct experiments all on your own, right? You are only seven." I said this with a grin, so they could see the implied challenge.

Of course the kids protested! "What?" they called. "We *can so* do experiments on our own."

"Are you sure?" I asked. "You'll have to write up a lab report on your own, following these steps." I pointed to the "To Write Like a Scientist" chart. "You really think you can design and record your own experiment?" I gave them each a new copy of another lab report booklet, again with four pages, each bearing space for a title on the top of the page.

◆ COACHING

In the connection, you often tie the new work to existing work. This will frequently involve referring to an anchor chart and adding one more bullet to that chart, one more strategy to children's repertoire. Today, instead, you call upon a broader sort of knowledge, suggesting that not only writers but also scientists revise. The principle is the same. You are relying on the fact that learning involves making connections, linking new information to existing information.

One of the challenges in a connection is to invite engagement and participation without losing control of the minilesson. You also want to teach efficiently and, above all, to save time for children to practice what you teach. This little bit of dialogue with the kids, and this just-for-fun jesting, are meant to support engagement, while allowing you to carry on briskly.

Name the teaching point.

"Today I want to teach you that scientists don't just follow someone else's recipe to do an experiment. No way! *Scientists come up with their own experiments!* They think, 'I wonder what would happen if . . .' and then they try it! Just like writers go through a writing process, scientists go through a scientific process."

TEACHING

Lay out some materials that kids can use when they devise their own innovations from the initial experiment.

"Since you will be revising our initial experiment, I brought in some materials that you could use." I brought out containers with blocks, different size toy cars, books, and a variety of math manipulatives. "These are the materials you can choose from to do your experiments today. You can also use the wooden blocks or big books as ramps. You and your partner will come up with one shared experiment that you will do together and that you will write about separately. When you do this, you will want to keep the materials in mind as you think about what you could change."

Demonstrate your step-by-step process: reread your lab report, think about how things could have gone differently, imagine a way to test things out, plan a new experiment, then record it.

"So, writers, I want to show you how a scientist might revise an experiment like the one we did yesterday so that you and your partner can make your own revisions on that experiment. First, I reread my initial lab report on my first experiment. As I do this, I recall that experiment and ask myself, 'I wonder what would happen if . . . ' When I ask that question, I am thinking about one thing I might change in that first experiment that might produce different results."

I read through the beginning of my lab report, "Hmm . . . the question was 'Will this little car go farther off the ramp on carpet or on bare floor?' My hypothesis was that the car would go farther on the bare floor because the floor is smoother than the carpet." I paused here to think aloud, "Well, my hypothesis was right. The car didn't go very far on the carpet. But maybe there's a way to make the car go further on the carpet. Hmm . . . let's see." I left a pool of silence, giving the students time to think along with me. "I wonder what would happen if I made the ramp totally flat—if it was all on the ground. How fast would the car go then?" I asked. "I think I am going to try that in my experiment today."

Teachers, you can be sure that I would far rather test out whether giving the ramp more height might alter the angle enough to accelerate the car's progress, but I am deliberately not going to choose to do the work that is most obvious for kids to do. I want to leave that for them to discover and try out themselves.

Recall what you did that you hope students do when conducting their very different experiments.

"Writers, did you notice how I reread my lab report and thought about what I had done and learned from the initial experiment? Then, I asked myself, 'I wonder what would happen if . . . ' and I came up with a way to revise the experiment and a question to test out."

ACTIVE ENGAGEMENT

Extract from students a recount of what they should do first, next, and then channel them to do those things. Coach into what they do.

"Now it's your turn. What do you and your partner do first?"

I easily could have told kids what they do first, but that would have been a missed opportunity for children to do intellectual work that stretches their thinking. By thinking about this, you are thinking about the DOK—the Depth of Knowledge—your teaching supports.

Children called out that they would reread their lab report and think about how things might have gone differently. I nodded, adding, "And you and your partner will want to think about what you learned from the experiment and come up with some ideas about how you could revise one thing in the experiment. Start rereading what you did before—with your partner, of course." After a few moments, I said, "Talk about your ideas. As you think about how you could alter the experiment, remember you can use only the materials we have. Plan together what you are thinking of doing."

I listened in as they shared their ideas. I heard some students say that they were changing the experiment completely. I stopped them to provide some coaching, "Writers, when scientists revise an experiment, they don't change everything at once. They choose one thing to change and test out to see how that affects the results. Talk with your partner again, and see if the two of you can come up with just *one* thing to change from the original experiment—like how I changed only the height of the ramp."

LINK

Channel writers to review the writing they will be doing during the various stages of the experiment. Get them started writing the first parts of their new lab reports while still in the meeting area.

"Writers, now that you have some ideas about how you will change the experiment, you will get to try those ideas. But before you get started, look over our chart that we have been keeping on writing like a scientist and remind yourself of the writing you will do before you start the experiment, during the experiment, and after the experiment." I left a little space for the children to do this thinking.

"The first thing you'll need to do is to figure out what the question is that you or your partner are asking, so that you can write that. You will have one shared question. If you can each come up with your own individual hypothesis, that would be great. Remember, that is a guess about how it will turn out. Then you and your partner need to decide the materials you'll need and your procedures. Will you get started with your partner making these decisions and doing all that writing, right here on the carpet?"

"Do that here on the carpet" makes sure that students don't bypass the writing altogether. As kids continue working in the meeting area, you can tap one partnership and then another, getting each partnership set up in a different part of the room, ready to try their versions of the experiment.

I gave students a few moments to write with their partners. "When you and your partner have written up your procedures, raise your hand, and I'll help you get the materials you need to get started."

Obviously, you can set up some children with materials and send them off to work before they are finished writing their procedure pages, allowing you to stagger the time when children begin and need your organizational help.

CONFERRING AND SMALL-GROUP WORK

Coaching Partners to Help Each Other

THERE WILL BE A BUZZ TODAY, because children will be excited to conduct their own experiments—and they will want to race quickly through half a dozen different ways to adapt the experiment. The reason that you ask them to write three portions of their lab report while still sitting in the meeting area is because otherwise they are sure to rush through these writing parts to get to the work with ramps and cars, discarding all that they have learned about writing like a scientist. The good thing is that as children work on the rug together, it will be easy for you to use voiceovers to remind them of all the work they did in their initial lab reports. You'll especially want to remind them to show all that they have learned about writing effective procedural pages.

If you pull up to a student who has forgotten to write with detail and clarity, enlist the help of her writing partner to encourage her to work on her writing. You might say to the writer, "Before I give you the materials for your new experiment, could you read what you wrote to your partner to make sure you have everything written down that is needed to complete the experiment?"

In instances such as this, it is always helpful to coach in to the ways that writing partners respond to each other. You can, for example, say to the writing partner, "Your job is to envision your partner's writing and to signal if the writing is confusing or unclear so she knows where she needs to go back and revise."

You might stay for just a minute, watching as the student reads her writing, and coaching, if necessary, to be sure that the partner tries to envision the steps that the writer is listing for the experiment. "Can you picture each step of that?" you can ask. Chances are good that your second-graders will need reminders to be explicit, detailed, and specific enough in their how-to writing that others can correctly follow their directions. Make sure, then, that partners notice when the writing is unclear and say, "Wait, I'm confused! What comes next?" or "Can you say more about that?"

Of course, part of your work will be helping *writers* realize that a partner's confusion is precious feedback, indeed. This doesn't just require them to orally elaborate, it also requires revision. Play up the value of revision and make the carpentry of this good fun. Writers can add flaps, inserts, new pages, and the like.

MID-WORKSHOP TEACHING

Multiple Tries and Detailed Records Matter

"Writers, can I stop you for a minute?" I waited until all eyes were on me. "Wow, wow, wow. You are remembering that scientists conduct multiple trials and checking to see if you get the same or similar results. My hat is off to you," I said, and swung an imaginary hat off my head, bowing low as I did so.

"Here's one tip. Don't forget to record as much as you can so you capture everything you remember from the experiment. That way if the results are surprising—say one time the car fell off the ramp altogether—then you can reread what you did (reread your procedures) and think, 'Did I do anything differently this time, even though I was aiming to do the same thing?' And others can reread your procedures as well, asking the same question.

"Take a second and look at your booklet to think about the amount of writing that you have completed. Ask yourself, 'Do I have enough information on the page to help me remember everything I did, with detail, in this experiment?' Right now, right here, set a goal that pushes you to do your most detailed writing. Then, write like your pencils are on fire!"

As Students Continue Working . . .

"Writers, I just want to remind you to look up here at our 'To Write Like a Scientist' chart. This chart can keep you on course. The chart can be like a checklist, helping you make sure that you do each part of your lab reports. Remember, if you are not sure what to include or what order to write your lab report, look up here!"

SHARE

Interpreting Scientific Results and Developing Conclusions

Select a partnership whose results didn't match those of the earlier experiment, creating a situation that begs for explanation. Then set that partnership up to share the experiment with the class.

"Writers, bring your latest booklets with you to the rug so we can talk about the new experiments you tested today."

As students gathered, they were intrigued by the scene that was set up. I had unstapled Brianna's lab report and taped up her pages on a blank sheet of chart paper. Brianna had a pointer in her hand as Sam read from his own booklet. They were poised and ready to teach.

"Scientists hold *symposiums*, which are like scientific conversations about latest studies. Imagine we are at a science symposium, and Brianna and Sam are on the podium, wearing their lab coats and teaching us about their experiment—what they first wondered, what they tried, and what happened. In a symposium, scientists often first walk listeners through their problem, hypothesis, procedure, and results, so let's let them do that."

Sam began, "So, first Brianna and I wondered, 'Would a snap cube travel farther on carpet or on bare floor?' Then we got snap cubes from the math center, and we still used the ramp, the bare floor, and the carpet."

"Brianna, can I give you a tip about what scientists do at symposiums? At a symposium, it helps to point to the objects or to drawings of them as they are mentioned, so people are hearing and seeing. Sam, could you start over, and this time, wait for Brianna to point to the objects in her book or to touch the ones on the table in front of her?" Sam repeated what he'd said, this time with Brianna providing the accompanying visuals.

Sam continued, "Brianna and I had different hypotheses. Brianna thought that the cube would go the same on both the carpet and the tile, but I thought it would go further on the tile, like the car. So then we tested it out and found out that it was hard for the snap cube to even go down the ramp, maybe because the ramp wasn't very high. We had to push it a little bit or it didn't even go. It was much easier for the car to do it."

Brianna added, pointing to her hypothesis page, "So we were both a little wrong. The cubes didn't even go down the ramp."

I jotted what the two children said onto a piece of paper:

> It was hard for the snap cubes to even go down the ramp. We had to push a little bit, and it didn't even go. It was much easier for the cars to do it.

Explain to the class that when writing the conclusion page, it is important to ask "why?" and to speculate about answers. Channel the whole class to do this work to make sense of the experiment the duo just shared.

"Class," I said "this is their result. They need to write that on the results page. But they also need to write their *conclusions.* So far we have studied how to make your procedure pages really good, but now I want to tell you that when you write the *conclusion* page in an experiment, it is important for you to ask the question—why?

"Why didn't the snap cube go down the ramp?" I wrote "*Why?*" onto the same paper underneath the children's result and placed the paper on the ground in the center of the meeting area. "Start by talking within small clusters—what do you think? What is a possible explanation for what has occurred?"

I then suggested we talk as a class and that Reagan get us started. "Maybe the wheels of the cars can go down a ramp easier than a cube that's flat," he said.

Jackson added, "Yeah, but a pencil *did* go down the ramp without any help from us." Then he added. "Wait, it's rounded. Like the car. The cube is flat, so it can't go without a push."

Sam jumped up. "We should have a test! Who did round and who did not round?"

I pitched in. "Sam, can you say that more clearly? I think you are on to something, and I want to be sure we are following your thoughts."

He tried again. "To see if it has to be round!" We waited and he continued. "We should tell our experiments and see if the only objects that went down the ramp are the round ones!"

Tell children that scientists often consult outside sources—other scientists' experiments, other articles and resources—to help them interpret their own results and write their conclusions.

"Scientists often need to know the results of other experiments, and they need to read background information about their kind of experiment, to help them write their own conclusions. Let's do some of that work starting right now, and we can continue if needed tomorrow and forever more. Take a moment to tell the people around you about the results of your own experiment and listen to their results—see if it changes your understanding of your own conclusion! Does it help you get some ideas about why your experiment turned out the way it did? Turn and talk."

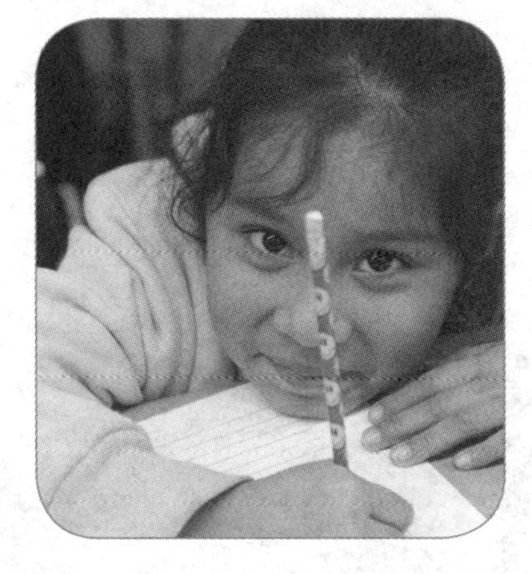

Close class by reminding students that their conclusion pages need to reflect that they have been asking why and developing hypotheses to explain what happens.

After children had had time to share results and take notes toward revisions, I said, "Scientists, you have been developing explanations for why things happened, and that is what scientists do in their conclusions. Those explanations—like ours—are often ideas that need to be tested out. Now you have come up with some new hypotheses to be tested! Anytime you have a hypothesis about something, you may want to try some new experiments and consult some other resources to revise and expand your conclusion. You might want to do this work after writing time or tonight."

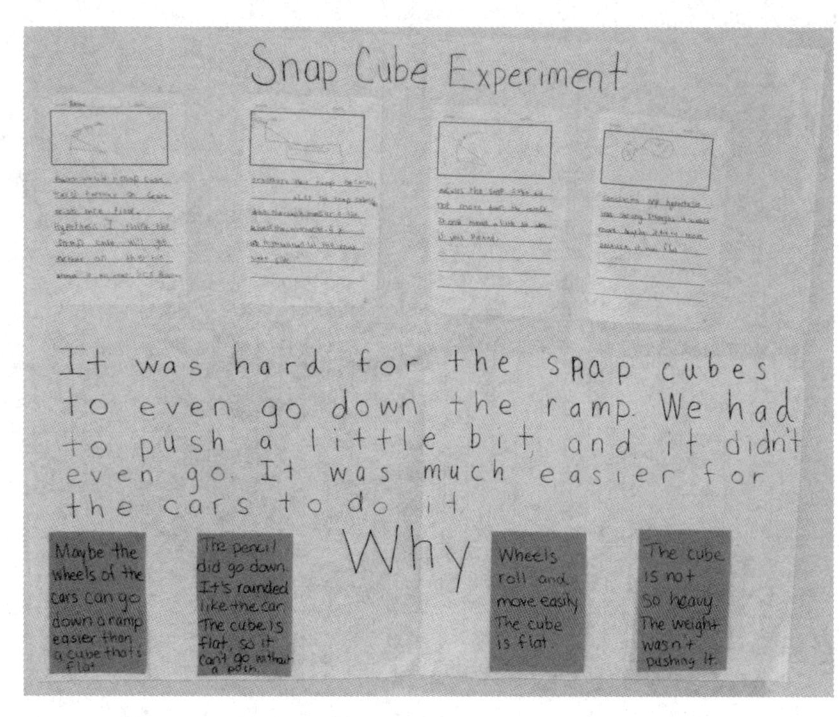

FIG. 3–1 The snap cube experiment

Session 4

Authors Share Scientific Ideas/Conclusions

THIS SESSION IS SIMILAR TO SESSION 2, in which you taught students to study a mentor text to learn about the characteristics of lab report procedure pages. This session brings children's attention to the results and the conclusion sections. It is hard to emphasize enough the importance of this part of a lab report because this is the part where researchers ask, So what? and What next? This is the time when researchers stop *doing* long enough to reflect. In the conclusion section of a lab report students shift from writing about concrete actions toward writing about ideas, from writing about what they *did* toward writing about what they *think*.

In this session, you will pose an inquiry question and invite students to learn more about informational writing. In first grade, your writers concluded their teaching books with a sentence. Today, you can say, "Now that you are second-graders, your conclusions cannot just be an ending sentence. They can, instead, be ending sections!" Tell them that they will not just recap how the event ended but also think about the event. Teach them to linger with their results, to highlight the most significant aspects of the work, to imagine next steps. In short—teach them to conclude.

This is a big deal. In all too many science classrooms, kids are involved in an experiment, they record results, and then the class ends, and the teacher calls, "Clean up!" It is essential that there be time for students to gather with each other and to talk about what they learned, what they will carry with them, what they want to continue to study. And that step backward is as important in a lab report as it is in a science classroom.

So today, you'll teach students, through the use of a mentor text, to conclude. This time, the mentor text will be a piece of writing done by a student that exemplifies the work you expect your writers to produce.

You'll invite children to revise their conclusion page—and then every other page as well. Later, you can spotlight ways to revise results as well. The bigger message is that writing is a process of drafting, and revision, or writing, and then pulling back to ask, "What am I learning?" Conclusions then become new beginnings.

IN THIS SESSION, you'll teach students that scientists spend a lot of time writing and thinking about their conclusions. They push themselves to ask Why? and then offer the best explanation they can based on their results. In this way, their conclusions often lead to more investigations and research.

GETTING READY

- ✓ Students' booklets of their experiments drafted in the previous session
- ✓ Prompts to support partner talk (I think the reason that the snap cubes didn't move . . . , Maybe it's because . . . , Or then again, maybe it's because . . .) written on chart paper (see Connection)
- ✓ Draft or student sample of a conclusion page to display *or* have ready copies of the page—one for each partnership (see Teaching and Active Engagement)
- ✓ "In Conclusions . . . " chart (see Teaching and Active Engagement and Mid-Workshop Teaching)
- ✓ Prompts to support writing circle talk about hypotheses (One reason is . . . Another reason is . . .) (see Conferring and Small-Group Work)
- ✓ Revision strips, flaps, pens, staplers, and tape ready for students in the writing center to revise their reports

COMMON CORE STATE STANDARDS: W.2.2, W.2.5, W.2.8, W.3.2.d, RI.2.6, RI.2.8, RI.3.8, SL.2.1, SL.2.3, SL.2.4, L.2.1, L.2.2, L.2.3

MINILESSON

Authors Share Scientific Ideas/Conclusions

CONNECTION

Remind children that the previous share session left them asking why, and channel them to continue speculating explanations for that phenomenon.

"In our last session you had the chance to try your own experiments with your partners and to think, talk, and write like scientists. You asked that all-important question: 'Why? Why would the snap cubes *not* roll down the ramp, when the flat stapler *did* go down the ramp?' Did any of you continue trying to figure out explanations, even after writing time was over?"

Many children signaled yes. "So, scientists, will you turn and share your thinking with your partner? Partner 1, say to Partner 2, '*I think* the reason that the snap cube didn't move down the ramp when the flat stapler did is that . . .' and then clearly spell out your hypothesis for this. Go."

Coach partners to challenge each other to speak with more clarity. Encourage listeners to try to follow the speakers' ideas.

The children talked, with many of them producing circuitous, confusing theories. Madison, for example, said, "The snap cubes didn't go down the ramp because the ramp is down and that's why it didn't go down, and the stapler did because it could go down the ramp."

After a little bit, I stopped the class. "Do you remember that yesterday Sam took a few tries before he was able to explain clearly the idea that we should investigate whether the difference in our results related to whether the objects we sent down the ramp were round? It isn't easy to put complicated ideas into words. So you need to be ready to say to each other, 'I'm not following what you are saying. Can you try saying that again in a different way?' You might even think of a way to help your partner say it. And then, you need to try hard to understand what your partner is saying. Turn and talk again."

◆ COACHING

Ideally in an instance like this it helps if you have written out the words "I think the reason that the snap cubes didn't move . . . " and "Maybe it's because . . . " "Or then again, maybe it's because . . . " It will take a fair amount of scaffolding for children to talk the talk of scientists, and everything they do orally is rehearsal for the writing you want them to learn to do. So don't think for a moment that the talk is not actually a form of writing instruction.

Accentuate the fact that scientists go through life asking, "Why?" Tell children that this kind of thinking goes into the conclusion of a lab report.

After children talked a bit longer, I jumped in. "We could all talk now about the possible reasons why some things do and some things don't go down the ramp—and we should keep thinking about that—but right now I also want to talk to you about the kind of thinking you are doing now. You are trying to figure out *why* things happen as they do. That is one of the most important things that scientists do.

"Scientists go through life asking, 'Why?' Have you ever noticed those roller coaster rides at amusement parks that chug up and up and up a hill so so slowly, almost come to a complete stop at the very top, and then zoom down that same hill with incredible speed? *Why*? Scientists notice and ask why. Have you ever noticed that when you are riding a bike uphill, really slowly, the bike starts to wobble—and when you ride it down hill, it goes straight as an arrow? Scientists ask . . . "

The children chimed in: "Why?!"

"Today, I thought we would talk about this sort of thinking because it goes on the conclusion page of your lab report. I thought we'd try to study some conclusion pages together."

It is true that we entered this unit unclear about what scientists do in their conclusion pages. This makes the question a true inquiry question. The other question on our minds was whether young kids could actually do the sorts of thinking that adults tend to do when they are concluding. For example, if kids draw conclusions that are not warranted in the data, can we teach them to discern this?

Name the question that will guide the inquiry.

"The question that we'll be researching today is this: When a scientist has collected some results and has formed new hypotheses about why she got those results, how does she write a conclusion?"

TEACHING AND ACTIVE ENGAGEMENT

Introduce a mentor lab report, and coach writers to research the piece as they read through it, learning how their own writing could go.

"Writers, here is a lab report from a second-grader, just like you! This writer did an experiment—not the same one as yours. After he did it, he wrote a page called 'Results,' where he listed which things floated and which sunk—like we listed how far things rolled, if they rolled at all.

"Then, he wrote a page called 'Conclusions.' We are going to study it not to understand his conclusion but to see how to write our own conclusions." I displayed a piece of student writing for everyone to see. Listen and read really closely so you can tell your partner what this writer does that you could try." I read the mentor piece aloud.

Notice that in this instance, the mentor text is addressing an entirely different research topic. It is often the case that when we want students to study how a text is written, we give them one that is off-topic.

Conclusions

My hypothesis about the experiment was wrong. I guessed that the paper clip would float and the wood would sink. But the paper clip sank to the bottom of the bowl! Why did it sink? Maybe because it is heavy even though it is small. Maybe it is made of silver and that makes it heavy. Maybe it has something to do with the shape of it? I know in the video we watched yesterday it said that the shape of something can make it push down into the water. And this connects to the paper clip because it is sort of skinny.

I was wrong not just about the paper clip but also the wood. I was surprised because it was big so I thought it would sink. But the piece of wood floated on top of the water in the bowl. I thought it was going to sink, but boy was I surprised! The wood is bigger than the paper clip but the wood floats and the paper clip does not. Maybe it's like what we learned about big boats and they still float. They have air inside–like a balloon, so they float. Maybe the wood has air inside. I don't know. It is a mystery.

If you feel like your students would benefit from a stronger mentor text, you may choose to write a lab report that you use in front of your students across several sessions, instead of using this kind of student example. You can make an informed decision based on the needs of your students.

Scaffold students' inquiry, collecting their observations on a class anchor chart.

"What did this writer do in his writing that you could try, too? Turn and talk. Tell your partner what you notice about this writer's conclusion page." After a couple of minutes I stopped the students and asked them to share.

Henry said, "He asked questions, like 'Why did it sink?' and then answered it, 'Maybe because it's heavy.'" I wrote that on chart paper.

Sara shared, "He told the reasons for why the paper clip sank and the wood floated. He didn't just give, like, one reason, he wrote, 'Maybe it's heavy or maybe because it is silver.'" I added that to the chart.

I added, "Reflect on your hypothesis" and put this first on our chart, as it appears first in the mentor text.

In Conclusions . . .

- Reflect on your hypothesis (My hypothesis was right/wrong . . .)
- Ask questions about your results (Why?)
- Give some POSSIBLE explanations– use ideas from other experiments and resources

LINK

Send students off to revise their lab reports, using all they have learned from the mentor lab report.

"Now that you've gathered some ideas from this mentor lab report, you will need to go back to your own lab reports to revise, using everything you know about writing like a scientist! You may want to go straight to your conclusion page to try out some of the things you noticed from the mentor piece today, or you may choose to revise any other part of your lab report. Think about what you would like to try in your writing today and get started! Once you've finished revising, I know you have more experiments to write up and conduct.

"Remember, whether you study a piece of writing or an experiment, it's important to study closely and learn from what you see. That will help you throughout your life!"

Obviously your chart will reflect what your children contribute—but on the other hand, you can always say that you also want to add a point and in that way get something you want onto the chart. You also steer this process by having children talk in pairs first, allowing you to decide whether their comments will be especially helpful.

CONFERRING AND SMALL-GROUP WORK

Using Revision Materials and Writing Partnerships to Bring Revision Work to Life

OVER THE COURSE OF THESE FIRST FEW SESSIONS, you have taught students several ways to revise their writing and their thoughts. You may also spruce up your writing center with attractive materials to entice your writers to take on this work, such as colored pens, revision strips, flaps, Post-it notes, mini-staplers, and tape. While it is not necessary to show every second-grader how to staple each and every strip onto the paper (remember, they just designed their very own experiments yesterday, and they have revised in these ways since kindergarten!), we do suggest you ask questions like, "What will you need to do that? Where might you add that in?" and then, "Reread that to see if it fits in there."

One of the important things to pay attention to in this session will be the quantity of writing children produce. In the lesson, you provided students with an opportunity to talk out their ideas with someone, and you may find it helps some writers to write a great deal more if you continue scaffolding their writing by offering talk time first. Be sure that

MID-WORKSHOP TEACHING **Conclusions Set the Stage for Further Investigations**

"Writers, are you noticing the same thing I'm noticing? When you make your conclusions, they end up sending you to try another experiment! For example, Claire was trying to think about why the stapler went down the ramp and the snap cube didn't, and she wrote in her conclusions, 'I think the stapler is heavy and the snap cube is light and the heavy stuff goes down more because the heavy stuff pulls it down.'

"And then, do you know what she did? She came to me and she said, 'I got another experiment I gotta do.' She wants to test out her hypothesis; she thinks maybe heavy things go down the ramp farther than light things. Turn and talk: How do you think she could test that?"

A moment later, kids had invented a whole raft of ideas. They'd said that she could put balls of clay on top of the car and see how far the car went with no ball of clay stuck to it, how far the car goes with a little ball of clay smushed onto its roof, and how far the car goes with a giant ball of clay on the roof. Then again, she could set up a whole lot of things like horses at the horse race, choosing things that are increasingly heavy, and then send one after another through the shoot.

"Writers, my bigger point to you is not about if something heavy goes farther down a ramp. No, my bigger point is that when you are writing up your conclusions really well and you grow ideas to explain what happened, it often happens that your explanation needs to be tested! So then you add a section called 'Further Investigations,'" and I added that to our chart. "And write about the plan—right there on the conclusion page. Then you go and get another lab report booklet, more materials—and you are off and running!"

In Conclusions . . .

- Reflect on your hypothesis (My hypothesis was right/wrong . . .)
- Ask questions about your results (Why?)
- Give some POSSIBLE explanations–use ideas from other experiments and resources
- Add further investigations

they don't just talk with you; writing partners need practice, talking and growing ideas together. This also frees you up to coach them and to move quickly on to other small groups.

You may decide to hold talking or writing circles where you coach a couple of writing partnerships as they share an explanation, discuss what it means, explaining their thinking, perhaps comparing the explanation to something they already know to illustrate their reasoning. ("The car went faster on the bare floor than on the carpet because the tiles on the floor are smooth and the carpet isn't. Maybe that means that things with wheels will go farther on smooth surfaces like hardwood floors, tiles, or plastic, than on things that are not smooth, like dirt, sand, or sidewalks. And maybe they will go faster on the smooth surfaces but slower on bumpy floors.") Then coach the writers in the circle to get all those ideas down on paper quickly.

Some students will need a bit more coaching to talk and to get down their ideas. You may decide to provide some prompts to help move these writers along. You might say, "Why do you think your hypothesis was correct? You might write, 'One reason is . . .' 'Another reason is . . .'" You will also want to direct them to the prompts they already know, "I think this happened because . . . " and "Maybe . . . or maybe . . . " Leave the prompts with the students, and remind them that they can use these and other prompts to help get their minds and pencils moving.

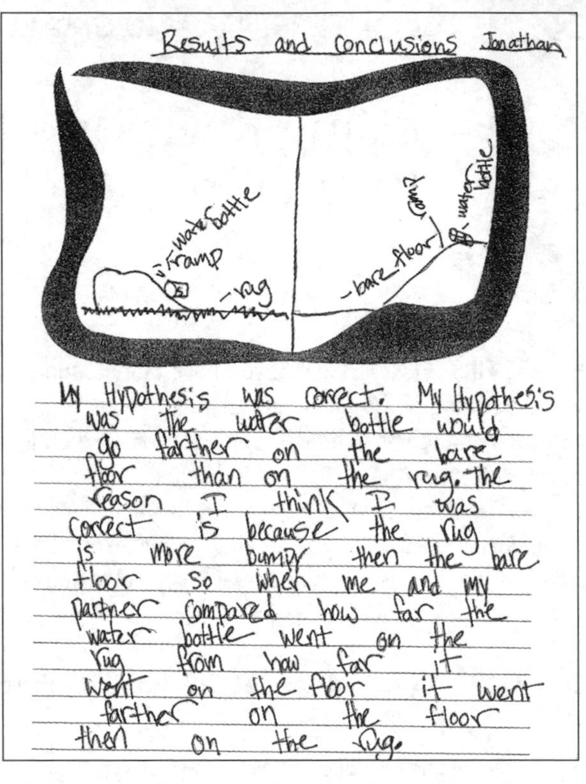

FIG. 4–1 Jonathan revised his conclusion to explain more by giving reasons.

SHARE

Connecting Science to Real-World Learning

Conclude the workshop by giving students a vision of how the work they did today could reach beyond the writing workshop, the classroom, and even beyond school.

"Scientists, this is the end of our workshop. I want you to think about everything you know about the experiments you did, especially the science behind it. I heard many of you mention the words *faster*, *slower*, *farther*, *moving*, *slowing down*. These words all go together under a *bigger* topic: forces and motion. Some of you may already know about forces or motion and what it has to do with your experiments. Turn to your partner and share what you know or are thinking about."

Henry turned to his partner, Connor, and waved his hands back and forth. "This is motion! Wave your hands like this!" Connor did, and Henry continued, "See how it makes wind? That's because there's movement!"

Stella turned to her partner, Jonathan, "I heard the word *force* before, but it means, like, when someone tries to make you do something. Like when someone forces you to jump in the water."

Jonathan responded, "Well . . . I've heard 'May the force be with you' before—that's from a movie. Maybe that means like the power or something?"

After another minute, I called the students' attention back to me. "I heard a lot of ideas as you talked about forces and motion. Some of you know a few things about forces and motion and some of you want to learn more, so here's what I want you to do between now and writing workshop tomorrow: find out more about forces and motion. If scientists don't know something, they just do some research and find out more.

"During recess today, you can ask some of your friends in other classes, 'Hey, do you know anything about forces and motion? What do you know?' Maybe when you go home, you'll ask an older brother or sister what they know about forces and motion. Or who knows? Perhaps you'll even look up the words *forces* and *motion* on the Internet!

"Movement, forces, and motion—they're all around us, so you can also be on the lookout for them. You can notice forces and motion at the playground, when your family is cooking dinner, or even at bath time! Think about what you know or are noticing. Tomorrow, we'll work on our experiments and our writing more and see if we can improve!"

Session 5

Scientists Learn from Other Sources as Well as from Experiments

IN THIS SESSION, you'll teach students that when scientists want to improve their writing, they learn more about what they are investigating. That is, scientists improve their writing by learning more science and then revise their writing based on what they've learned.

GETTING READY

- ✓ Clipboards (or something to write on), a stack of six to eight Post-it notes, and a writing tool
- ✓ Your own text/mini-lecture to read aloud (see Teaching and Share)
- ✓ Chart with scientific vocabulary words extracted from the read-aloud/mini-lecture (*forces, push/pull, gravity, motion*) (see Active Engagement)
- ✓ Forces and motion texts—a few leveled books, photocopied articles, photographs with captions and labels—set around writing spots for students to access other sources in the midst of writing (see Link), such as these:

  Forces and Motion, by John Graham (2013)

  Do-4U the Robot Experiences Forces and Motion by Mark Weakland (2012)

  Motion Push and Pull, Fast and Slow, by Darlene R. Stille (2006)

  Roll, Slope and Slide: A Book About Ramps, by Michael Dahl (2002)

  Inclined Planes, by Valerie Bodden (2011)

  Forces Make Things Move by Kimberly Brubaker Bradley (2005)

  Move It! Motion, Forces, and You, by Adrienne Mason (2005)

Common Core State Standards: W.2.2, W.2.7, W.2.8, W.3.8, RI.2.1, RI.2.2, RI.2.3, RI.2.6, RI.2.7, RI.3.1, RI.3.2, RI.3.3, SL.2.1, SL.2.2, SL.2.3, SL.3.3, L.2.1, L.2.2, L.2.3, L.2.4.a.d, L.2.6, L.3.6

ALTHOUGH YOUR FOCUS has been on writing lab reports, the unit itself is on information writing—and chances are that the one thing your students' work most needs right now is more information. It is impossible to generate information writing without information: numbers, facts, definitions, concepts. Certainly some of the information that students need is available from their research. They know the specific materials and procedures used and the data from their successive trials. But your students also need to know the scientific terms, concepts, and ideas that can illuminate their work.

The goal of today's session is to supply your students with enough scientific content that they can experience that special electric charge that happens when all of a sudden, one discovers new information. Eureka! It's like redecorating after receiving a truckload of beautiful new furniture. Look how much better everything looks now!

Today, to deliver that truckload of new stuff, we suggest you give students a mini-lecture. This is a form of instruction that we use a lot when teaching in social studies and science classrooms. By teaching kids to learn from a very tiny, compact lecture, you go a long way toward teaching them to learn from listening to or reading an expository text. A mini-lecture is just a text that has been tailor-made for them—and in this instance, for the precise work they have been doing as experimenters. You can, of course, design your own mini-lecture or you can borrow (and improve upon) the one that we gave to our students and include here. Either way, this will be a short, focused lecture in which you impart specific content that you want students to know and use. Your lecture should convey some domain-specific vocabulary and also help children grasp some key scientific concepts—above all, ones that illuminate the work children have been doing with forces and motion. For example, you'll convey that information exists in the world that can explain why a little plastic cube might sit on the top of a gently elevated ramp.

Of course, you will also be teaching students the skills they need to listen to a lecture. You can draw on all the work the class has done around nonfiction read-alouds. Remind your students, "You already know how to listen, think, and jot during read-alouds, so you will do the same today!" Then direct their thinking and jots to the relevant content that will help them understand their experiments. "Writers, listen with your ears wide open to learn as much scientific information as you can." If you want your children to take notes, set them up with three or four Post-it notes and encourage them to write one note on each. These can then be transferred easily to students' writing.

"It is impossible to generate information writing without information: numbers, facts, definitions, concepts."

After the mini-lecture, your second-graders will go back into their lab reports, searching for places to incorporate the new information. As they do this work, encourage them to thoughtfully consider where to add the content, "Where can you put that special scientific word? Does that make sense here?" Your students may do this work a little haphazardly at first, but the main goal is that they are thinking about the content and trying it out in their writing.

MINILESSON

Scientists Learn from Other Sources as Well as from Experiments

CONNECTION

Channel children to share what they know about what scientists do, then suggest today you will add one more item to their list.

I called the students over to the meeting area. "Bring clipboards, pencils, and a stack of Post-it notes," I said. Once they'd gathered, I continued. "Writers, can you name three important things you've learned so far about how scientists learn and write? Turn and tell your partner." I listened over the hubbub.

The students noted many different things—that scientists write a hypothesis, they write everything they do in an experiment, they try the experiment over and over, and they use scientific words. "You know a ton about how scientists go about writing lab reports. Today I want to add one more thing to all that you already know."

Name the teaching point.

"Today I want to teach you that the more a person knows about a topic, the better he or she can write. Sometimes when you want to improve your writing—say, you're writing about cars and snap cubes and staplers sliding down ramps—the best way to improve the writing is to learn more about how the world works. That is, sometimes the best way to improve your scientific writing is to learn more science."

TEACHING

Elevate the idea of learning from a lecture by suggesting this occurs at colleges all the time. Explain that you will give your lecture twice and set children up to take notes.

"There are lots of ways to learn about a topic. A person can read a book, or talk to an expert, or watch a video. At college, people listen to a certain kind of speech on the topic. If a speech is *full* of information, it is sometimes called a *lecture*.

◆ COACHING

This is a very simple connection. You choose this because the rest of the minilesson is complex and rich. So you essentially create some throat-clearing time. This is a way to ask, "Ready?"

"Today I am going to give a mini-lecture about some scientific concepts that will help you think much more about the work you have been doing with ramps and motion. Your job will be to learn in ways that let you teach other people this new information. That is what you do in your lab report. I'm going to use gestures and actions that will help you learn, and you might use them as you listen, too. And again later, when you teach.

"This mini-lecture is about forces and motion, a topic we have been thinking about since the beginning of this unit. (Actually, though we may not have known it, we've been learning about it for a lot longer than that!) I'm going to give the mini-lecture twice. The first time, listen and get ready to reteach it to your partner. The second time, I'll ask you to take some notes on Post-its. Are you ready? Gesture with me to help yourself take it in."

Both mini-lectures are excerpted from http://www.ehow.com/info_8580833_motion-force-kids.html by Asa Jomard.

Forces

Things that you cannot see are exciting but also annoying, and it is easy to describe them as mysterious. Forces *may be invisible, but you can see the result of forces every day. Several forces are acting upon you right now, and a force is* a push *and* a pull.

"Okay, are you picturing what I just said? A force is a push and a pull. Hmm . . . " I said as I swayed back and forth lifting up my hands as I pressed against them.

When you go on a roller coaster, your feelings are caused by a force that pushes your stomach upward when you go downhill. When the train changes direction your stomach is pushed downward. The invisible forces make you giggle and scream.

"Put yourself on that roller coaster. Are you holding on tightly?" I said as I gestured the movement and scanned the class to watch the students assume the position. "Now, I am going to say this part again. Put yourself on the roller coaster, and think about how your body changes and feels when you are going down." I started to lift my body in a clenched position to resist the force. The students began to do the same.

Motion

Everything moves, but some things move very little or slowly. Objects tend to resist changing their state of motion. *Things that stand still want to stay still, and things that are moving want to continue to move. If something is going to move a force needs to be involved. Some movements happen in a straight line, but some have a circular movement. When you throw a ball through the air there are forces pushing the ball from different angles. That is why a ball goes straight or curves in a different direction after it is thrown.*

"Hmm . . . let's think about that for a moment and try to picture this. We need to stop to ask ourselves, 'What does this mean?' Then, we can try to think of an example. What is still and wants to stay still? Hmm . . . a big rock does not move by itself. What is something that is moving and doesn't want to stop? Hmm . . . a rolling ping-pong ball doesn't want to stop."

As you prepare for this lesson, you may want to read the mini-lecture through the lens of determining importance. You might try underlining important parts to cue yourself to read a particular part slowly and with emphasis. This can support students' listening comprehension.

You may decide that your students need a mini-lecture with slightly different content. There are several sources to choose from when creating a mini-lecture. For example, you might choose an excerpt from a read-aloud text or an article you find online, or you may decide to draft your own mini-lecture that meets the needs of your particular students.

ACTIVE ENGAGEMENT

Ask students to turn and teach each other what they just learned.

"Partner 1, you are a Professor of Forces and Motion, and it is time for you to teach a class. Turn to your partner and teach your partner everything you have learned about forces and motion."

I listened as the children talked. "Professors, don't forget to use gestures, actions, and examples to make your point. I love the way Sophie is throwing an imaginary ball and showing how the air, pushing back against that ball, makes it change direction! Let me see the rest of you using gestures, actions, and examples. I've got some words up here on chart paper that you may want to include in your lectures as well," I said, and pointed to the chart paper where I'd written:

force
push
pull
motion

"Now go ahead and teach!"

Return to your lecture, and this time channel students to listen and take notes in ways that prepare them to talk about their experiments in forces and motion. Then get them talking.

"Now, I'm going to give the mini-lecture a second time. This time, will you take some notes about some of the parts that make you feel like there is a connection to your experiment? Write down each thought or each part you want to remember on a separate Post-it note. Then, you'll be able to stick that note onto the part of your lab report that you think is related. Ready?" I read the first part of the article again.

"Who has a note they can share?"

Jonathan read his note, "Forces are invisible. If something moves a force is involved."

"Where do you think we could stick this note in one of our lab reports? Where is a connection from this information to this experiment? Where in this experiment might an invisible force be involved? Talk to your partner." After a moment, I asked for ideas.

Sal said, "This makes me think that the car went down the ramp because of an invisible force. We could put the note on the part about the car going down the ramp." We all agreed so I asked, "Now we have made a connection. How can I explain that in a conclusion? Tell your partner."

We find it helpful to deliver the same lecture twice so that students have multiple opportunities to process the information presented. We also find it helpful to have them process out loud with each other before asking them to take written notes.

LINK

Set children up to read more sources and to take notes about new information to then add into their writing.

"You can continue doing this work yourself—you can take notes from outside sources and use them to help you get ideas about why your experiment happened the way it did. In other words, now that you have all this information about forces and motion, you can go back to your writing and find places where your new knowledge will help you revise. You might want to start by sticking them on the lab report, but then see if you can revise what you wrote so you say a lot about what you have been learning. Off you go!"

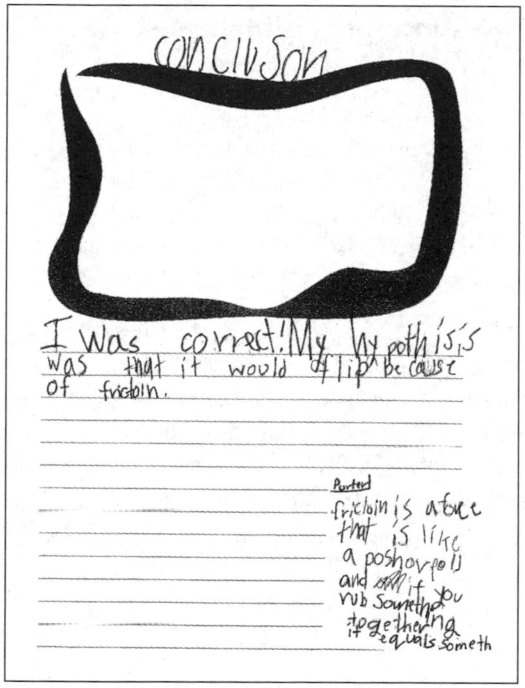

FIG. 5–1 Porter's notes from the mini-lecture

Supporting Writers' Learning Trajectories

ONE OF THE THINGS you will want to remember is that you are never just teaching the content of that day's minilesson. Your conferring and small-group time is your time to continue to give students feedback on their progressions along trajectories that have relatively little to do with the specific unit. And so although your unit is science writing, you are still teaching students to write with volume, to use conventions with automaticity, to plan their writing and revise those plans even before starting to work, and so forth. And during the research phase of any conference, you will want to call to mind the many agendas that are at play in any one moment.

For example, if you sit next to a student and watch her work, although you will scan her writing and try to determine what she is attempting to do as a writer, you will also want to scan your notes from the last conference to remind yourself of the contract you and the writer established then. This can help you angle your research question. Instead of starting with the open ended, "What are you working on?" you are able to say, "It looks like you are revising your writing. I recall the last time we talked, you agreed to add more paper so that your revisions wouldn't just be one word here or there. Can you walk me through the revision you are doing now and tell me how it has been going, trying to add larger revisions?"

MID-WORKSHOP TEACHING **Incorporate Information and Technical Vocabulary into Writing**

"Scientists, you should be using the information you learned on almost every page of your lab report. How many of you have added some of the information you learned into your diagrams?—thumbs up. This will make a *huge* difference in your conclusions. You'll probably end up needing to write big flaps of new information or whole new pages!

"Scientists, whenever you try to learn about a new topic, you should pay attention to the special words that go with that topic. Like if a new kid came to our class and was trying to learn about writing workshop, the kid might pay attention to a word like *minilesson* or *revision*. Make sure that you are using some of the new vocabulary in your writing. Think of the important words you learned to make your writing sound like an expert scientific report."

The interesting thing with this unit is that while your children are collecting and studying data from multiple trials with a car on a ramp, you will be collecting and studying data related to multiple trials involving your students and their writing. Just as sometimes one factor emerges as important when working with ramps, so, too, sometimes one factor will emerge as important when dealing with writers. Our sense is that this unit can allow you to see the impact that motivation can have on a child's writing development. We suspect that you have opportunities during this unit to turn around reluctant writers, renaming them as nonfiction writers, as scientific writers. So absolutely, study the patterns in your writers' development, and use this unit as a time to blast through the usual constraints. If a child usually revises in very modest ways, tap into the power of information to ignite whole new possibilities for revision. And then let the writer see how she is changing, and believe those changes are for good.

Using Sources for More Information

Challenge students to consider how new information about forces and motion from another source influences their thinking about their experiments.

"Scientists, I'm going to share a little more information about forces and motion. As you listen, think about how you can use the information to think differently about your experiments and your partners' experiments. Be thinking about the experiments we've been doing, and how any of this relates in any way to what you see. Are you ready?"

Friction

When you are riding in a car and slam on the brakes, the rubber tires grip the concrete to help the car stop. The car stops because of friction. *Whenever any two objects rub together there is a force that slows them down. This force is friction. Rough surfaces, like concrete, cause more friction than smooth surfaces, like ice. Different objects experience different amounts of friction depending on how smooth, rough, hard, or soft they are. Any time one object slides or rolls against another, there is friction.*

Ask students to think and talk about what they will notice in their lives in light of the new information.

"So right now—will you turn and talk? What will you notice and think about on your way home today?" I listened in as the students talked with their partners.

Sara said, "I'm going to look at the train on the tracks. When I take the F train home sometimes I see the other trains and they rub on the track and make sparks!"

"That's cool!" Connor responded. "When I go home, I'm going to try rubbing two sticks together to make fire!"

Finley said, "The bus and the cars on the street! There's friction when the wheels roll on the street!"

"I think that shuffling your feet on the ground makes friction. Like how my shoes rub against the sidewalk, that's kind of like when wheels are on the street because they're both made of rubber," Coby added.

Luca said, "Sliding down the slide. When I slide it's kind of slippery, but sometimes I get stuck. Is that friction?"

I wrapped up the session. "Wow! I love the way you're thinking about where you can find friction in the world! Isn't it exciting how the science and new concepts we have been learning apply to so many different things in our world?!"

Session 6

Student Self-Assessment and Plans

IN THIS SESSION, you'll teach students that writers self-assess, making sure their writing reflects all they know how to do. Then they set goals for themselves, making plans to improve as writers of informational texts.

GETTING READY

- Your own model lab report to use in the demonstration (see Teaching); you will want to display your writing on an overhead or have copies for students, one per partnership
- Enlarged copy of the Information Writing Checklist, Grades 2 and 3, to use during teaching
- A pointer to engage students in a shared reading of the checklist (see Teaching)
- Copies of the Information Writing Checklist, Grades 2 and 3 to be distributed during active engagement—one for each partnership and extra copies to leave at tables
- Mentor texts that students can read and emulate as they set writing goals (see Conferring and Small-Group Work)
- "Words Science Experts Use" chart (see Mid-Workshop Teaching)

Common Core State Standards: W.2.2, W.2.5, W.2.6, RI.2.1, RI.2.10, SL.2.1, SL.2.3, L.2.1, L.2.2, L.2.3, L.2.6, L.3.6

AS YOU COME TO THE END of the first bend of this unit, your students' confidence and ability to write like scientists have grown. Early on in the bend, you asked your students to assess their own writing, using the Information Writing Checklist for grades 2 and 3. In this session, you will once again ask your students to self-assess, holding their writing up to the Information Writing Checklist, Grades 2 and 3.

The goal of this session, then, is to support your second-graders in becoming more reflective and independent writers, instilling confidence in them so they are able to take on the challenge of checking their own work against a second- and third-grade checklist, to reflect on what they are already doing and what they have yet to try, and to develop goals based on this information. In this session, you will build excitement around the checklist, proclaiming that your students are ready to be critical readers of their own writing, and reinforcing the idea that writers are always finding ways to improve and build on what they know. Remind your second-graders that as they learn more about writing informational texts, and about writing in general, they can stop and reflect on this knowledge, develop next steps, and create a vision for what their writing should look and sound like at the end of the unit. This will help them to carry on with independence, both now and throughout their writing lives.

At the end of today's workshop, you will "publish" students' lab reports by sharing them with another class. You have been telling students all along that readers of their lab reports need to be able to replicate their steps exactly in order to try out the same experiment and get similar results. By having children send their lab reports out into the world—even to another class of second- or third-graders—you will reinforce the idea that all writing—even scientific writing—has an audience.

MINILESSON

Student Self-Assessment and Plans

CONNECTION

Build energy for this new session by challenging children to be as independent with checking their writing as they are with other second-grade routines.

"This morning, when you were all walking into the room and unpacking, I remembered back to the beginning of the year when we had to practice our morning routines over and over, and some of you needed reminders to take your notebooks out of your backpack, or to hang your backpacks in the closet, or to turn in your homework. But now, you can do the whole routine without my help! You came in quietly, unpacked your backpacks, put your notebooks and folders away, and turned in your homework all without me having to tell you what to do. You don't need me to remind you or check your backpacks to make sure you unpacked everything, right? You're grown-up second-graders who can do that all on your own!

"Well, that got me thinking that you are ready to do lots of things without my help. You already know a lot about writing and getting your writing ready to share with others, and I bet you are ready to check your writing on your own, just like you do the morning routine on your own. You don't need me to check your work for you because you can check your own work! Are you ready for this challenge?"

Name the teaching point.

"Today I want to teach you that second-grade writers can figure out how to make their writing the best it can be. You can use the Information Writing Checklist to help you. You can read the checklist, then go back to your writing to see if you did these things. Once you have gone through the checklist, you can look at the items that you have not checked off and make writing goals for yourself."

TEACHING

Build excitement around the second- and third-grade checklist.

"Remember at the beginning of our unit when we looked at the Information Writing Checklist to make sure you were doing everything that you know to do and to look forward at what you will be working toward in this unit?" The students nodded. "Well, you already learned so much about information writing in this unit that you are ready to check

◆ COACHING

It's always helpful to reinforce what students know and can do before recruiting them to try something new. This is a simple minilesson that makes a very obvious point. At the start of the year, there were so many classroom routines that were new to the children—now these are second nature to them. In the same way, you will soon say, learning to conduct experiments and to write lab reports will soon be second nature to the children as well.

your work against this checklist again. We are not even close to the *end* of second grade, but some of you are doing a lot of the things on the second-grade checklist and some of you are even doing things on the third-grade checklist. Are you ready to check your writing again using these checklists?" The Information Writing Checklist, Grades 2 and 3 can be found on the CD-ROM.

As students exclaimed "Yes!" I flipped over my chart paper to reveal the Information Writing Checklist.

"Let's read over the checklist together."

Use a pointer and engage in shared reading, if you want. Perhaps you'll want to vary this and read some parts to them, just to keep kids' attentiveness high.

Demonstrate using the checklist with your demonstration lab report and setting goals for upcoming work.

After reading the checklist I pulled out my own lab report. "Writers, let's try out this checklist with my writing. I'm going to go through the first section of this checklist that says 'Structure.' Watch me as I read the checklist, item by item, then check my writing to see if I did these things.

"Let's see, the first item is 'I taught readers some important points about a subject.'" I took a moment to look at my report. "I can't check that item off until I find that part in my writing. Let me see—yes. I did that." I read that part aloud. "So, I'll check that item off!

"Next, I'll look over at the third-grade column. The first item there is 'I taught readers information about a subject. I put in ideas, observations, and questions.'" I looked at my report again. "I did teach readers information about my subject—but what does, 'I put in ideas, observations, and questions' mean? Hmm. I'm not too sure I have done that or that I even know what that is, so I'm going to make a star here. How exciting—I have a star to reach for!

"Next item, 'I wrote a beginning in which I named a subject and tried to interest readers.' Hmm . . . No! I didn't do this *yet*, but I can soon. I did name my subject, but I'm not sure that I tried to interest my reader. I'm going to star that part. And, looking over the third-grade column, I'm also not sure I got readers ready to learn, so I'll star that too.

Make sure you demonstrate being critical and identifying upcoming work. That is your goal for students, too.

"Okay, next item, 'I used words like *and*, and *also*, to show I had more to say' Hmm . . . Yes! I have those words here and here," I pointed to my writing, "so I can check that off!

"Okay, I'm going to continue going through the checklist later. But my next step will be to think about the items that I starred: 'I put in ideas, observations, and questions'; 'I named a subject and tried to interest readers'; and 'I got readers ready to learn.' So, those are things I will try next time. I need to make a writing goal for myself so I can remember to work on those three goals.

"Let me look closely at my writing and think what I can really do. I'm going to study my beginning and get some help on how to put in ideas, observations, and questions. And I am going to make a goal for myself to interest my readers and get them ready to learn."

Restate the transferable strategy.

"So, writers, did you see how I read the checklist, item by item, looked to see if I did those things in my writing, and then checked off the item if I did it, or starred it if I didn't? Then I made a goal for the items I starred."

ACTIVE ENGAGEMENT

Set children up to practice using another part of the second- and third-grade checklist.

"Now, do you think you can help me with the next part of the checklist, the 'Development' and 'Language Conventions' sections? I'm going to give you a copy of the checklist, and you're going to look at my writing and check off or star each of the ones you see in my writing. Ready?" Quickly, I dispersed one checklist to each partnership.

"Now, look at the items that you starred and come up with a writing goal for me. When you make a writing goal, think of what a next step would be to make the writing stronger. Turn and tell your partner a goal you would set for me based on the checklist and based on what I've done or haven't done in my writing."

Information Writing Checklist

	Grade 2	NOT YET	STARTING TO	YES!	Grade 3	NOT YET	STARTING TO	YES!
	Structure				**Structure**			
Overall	I taught readers some important points about a subject.	☐	☐	☐	I taught readers information about a subject. I put in ideas, observations, and questions.	☐	☐	☐
Lead	I wrote a beginning in which I named a subject and tried to interest readers.	☐	☐	☐	I wrote a beginning in which I got readers ready to learn a lot of information about the subject.	☐	☐	☐
Transitions	I used words such as *and* and *also* to show I had more to say.	☐	☐	☐	I used words to show sequence such as *before, after, then,* and *later.* I also used words to show what didn't fit such as *however* and *but.*	☐	☐	☐
Ending	I wrote some sentences or a section at the end to wrap up my piece.	☐	☐	☐	I wrote an ending that drew conclusions, asked questions, or suggested ways readers might respond.	☐	☐	☐
Organization	My writing had different parts. Each part told different information about the topic.	☐	☐	☐	I grouped my information into parts. Each part was mostly about one thing that connected to my big topic.	☐	☐	☐
	Development				**Development**			
Elaboration	I used different kinds of information in my writing such as facts, definitions, details, steps, and tips.	☐	☐	☐	I wrote facts, definitions, details, and observations about my topic and explained some of them.	☐	☐	☐
Craft	I tried to include the words that showed I'm an expert on the topic.	☐	☐	☐	I chose expert words to teach readers a lot about the subject. I taught information in a way to interest readers. I may have used drawings, captions, or diagrams.	☐	☐	☐

Gather students and reiterate comments students made regarding goal setting.

After a few minutes, I called the students back. "I heard some really good goals for me. Emily said that I could add at least one expert word on each page. Ryan said that I could include more science facts in my results page. That is such smart goal-setting! Now I know how to make my piece even stronger!"

LINK

Restate the teaching point, and set children up for independent work.

"Writers, as you go off today, each of you will have a copy of the Information Writing checklist." I held up the checklist once again. "I put extra copies at your table so there are plenty to go around. Remember, you might start by rereading the checklist and rereading your writing. As you reread, when you find something you've done, you can either check off the item or star it. Then, you can set your goals based on any you *haven't* yet done, and get right to work. Remember, using checklists like this one can be helpful any day, not just today. Second-graders, who is ready?" Immediately, thumbs shot in the air. "Off you go!"

CONFERRING AND SMALL-GROUP WORK

Supporting Writers to Turn Plans into Realities

YOUR STUDENTS WILL LEAVE YOUR MINILESSON today excited to take on the goals of the Information Writing checklist. You may want to take the first minute or two to assess how your students are using this new tool. As you scan the class, you may notice students are progressing in a few typical ways, and you might group them accordingly for small-group work. One group of students may be using the checklist easily, needing little to no support. Another group may be studying the checklist, a bit unsure how to proceed. The final group may be checking off items randomly, not really giving thought to each one. Each category of students could be grouped together for a small-group strategy lesson.

When you confer with the small group of students who are struggling to use the checklist—perhaps unsure what to do next after checking off a few items—you'll support them with problem-solving strategies by teaching through guided practice or inquiry.

As you meet with them, you can say, "Today, we are not going to wait until next time to fix our writing. This is what I want you to try. Put your finger on one place in your checklist where you have not checked off an item, or where you've starred something." Scan the group as kids make decisions and support students as needed. "Now I want you to think, 'Is there a book that could give me tips to help my writing?'" Allow students a minute to jump up and find a suitable text. A big part of what you are teaching the group is to find their own mentor texts. You want to teach a transferable strategy that the group can use whenever they are writing.

It is a good idea to have a mentor text or two by your side, however, to suggest to any students who are stumped and can't come up with a mentor text. Open up the mentor text to a page that students could inquire about and ask, "What is this writer doing that you could try, too?" The page you select should spotlight work outlined in the goals set by students in the group so that it sparks ideas to incorporate into their writing. Coach the children to find the materials and resources necessary to problem solve and change the starred items on their lists into checked off items.

MID-WORKSHOP TEACHING

Revising to Use the Words Scientists Use

"Writers, there is something really important that I want to talk to you about. I was looking at our checklist and noticed the section that said, 'I tried to include the words that showed I'm an expert on the subject.' It got me thinking that you now know many words that science experts use. Quickly, look through your lab reports with your partner and think about the words that you used to make yourself sound like an expert."

As students searched their lab reports pointing and naming scientific words they used, I began to make a chart entitled, "Words Science Experts Use." As I listened in, I wrote the following words: *hypothesis, trial, procedure, results, observe, compare, measure, data, force, motion, friction*, and *results.*

"Boys and girls, you are using all of these words in your writing." I pointed to our list as I reread them aloud. "Are you using our science words in your writing some of the time or all of the time? Right now, read through your writing and try to find a place where you could change an ordinary word into an extraordinary science word. Let's try to make your writing sound more scholarly and scientific all of the time. Are you ready? Go!"

SHARE

Goal Setting and Publishing

Set children up to share goals they have reached with a student who is aiming to do the same thing. Encourage them to use precise language to explain what they did.

"Today, all of you set goals for your writing—some of you wanted to use more expert words, some of you decided to improve how you introduced your topic to your reader, and others of you were working on making sure you used capital letters for names. You have been so focused today—and it's clear you've achieved some of your goals already.

"Right now, writers, look at your work and find one place where you already reached one of your goals. Put your hand up when you find it."

Once all hands were up, I said, "Great. What I'd like you to do now is to share what you did with one of your fellow writers. Share it with someone who is trying to reach the same goal you reached today. As you share, be as precise as you can. Show the place in your writing where you made changes and explain what you did to achieve your goal. When you sit down together, you will be swapping success stories. Each of you should be able to see an example of something on the checklist that you want to do but have not done yet. Give each other your expert advice. Off you go!"

End by "publishing" children's writing, sending their lab reports off to another classroom for other students to replicate and compare results.

"Writers, I'm glad to hear you celebrating the goals you have achieved and also to hear you setting new goals for yourselves. That is just what professional writers do. Since you have worked so hard on revising your lab reports, I thought we would send them out into the community—to Mr. Feeney's class or maybe even up to the third-grade class, too—to see if other students can replicate your experiments. Won't it be exciting to see what results other students get when they follow your lab reports? I know we didn't fancy up your writing completely yet, but tomorrow we will be starting a new experiment, so let's send these lab reports out into the world today. Congratulations, writers."

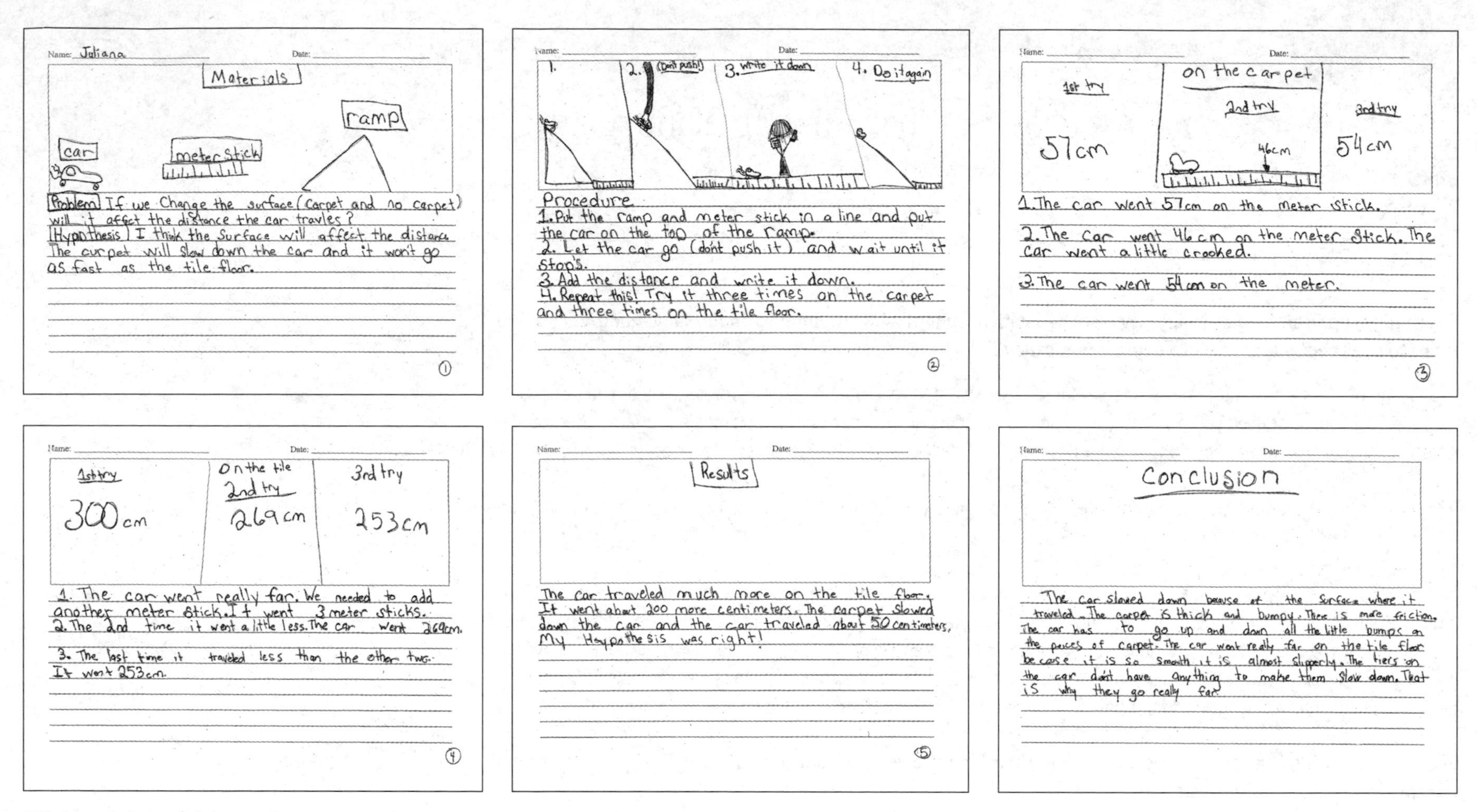

FIG. 6–1 Julianna's lab report

Writing to Teach Others about Our Discoveries

BEND II

Session 7

Remember All You Know about Science and about Scientific Writing for New Experiments

IN THIS SESSION, you'll teach students that scientists bring all they know about writing and about science to new experiments, drawing on all their knowledge to write well and conduct precise and replicable experiments.

GETTING READY

- "To Write Like a Scientist" anchor chart (see Connection)
- "In Procedures" and "In Conclusions" anchor charts, for students to reference
- Classroom spots for each pair of students to set up catapult experiments. We suggest providing partnerships with carpet spaces or cushioned landing surfaces to prevent the ping-pong balls from rolling too much after they land; this will ensure more accurate distance readings.
- Baggie of supplies for each partnership that includes a ruler, a plastic spoon, a rubber band, masking tape, a ping-pong ball, and a cotton ball; these can be kept at the meeting area—you may use one bag in the fishbowl (see Teaching and Active Engagement) and then distribute the remaining bags to each partnership after the Link.
- Several meter or yardsticks, ideally one for each partnership, to measure the distances the objects fly
- Blank booklet, a writing tool, and a clipboard
- Marker and dry erase board where you will post today's scientific question (see Teaching and Active Engagement)

Common Core State Standards: W.2.2, W.2.7, W.3.2.a.b, W.3.7, RI.2.4, RI.2.7, RI.3.4, SL.2.1, SL.2.2, SL.3.1, L.2.1, L.2.2, L.2.3

IN THE FIRST BEND OF THIS UNIT, you invited your students to write—and think—like scientists, asking questions, posing hypotheses, testing their ideas, and recording their findings. You introduced the lab report format, guided students to use mentor texts, and asked them to recall all they know about procedural writing to get a feel for how lab reports go. Along the way, your students have probably wanted to investigate the kinds of questions that grown-up scientists wouldn't (do ping-pong balls respond to whispered encouragement?)—and you have let them, in the spirit of scientific inquiry. Now, as you move into the second bend of the unit, your goal will be to help students begin to internalize the scientific procedures and writing processes they met in Bend I. By the end of this bend, your students should feel like they are experts at writing lab reports. Like learning to ride bikes without training wheels, they will gradually write lab reports with less scaffolding. They will pick up their booklets and get started recording important information as they experiment, they will write more detailed procedures, and they will write volumes after their experiment is over.

In this session, students will again be conducting experiments with their partners, but you will coach them to recall the steps for setting up and writing about those experiments so that they are working with greater independence. You will ignite students' enthusiasm for the new round of investigation by reminding them that by publishing their results they have become real scientists, joining the scientific community of their school. As such—you might tell them—it will be essential that they communicate clearly all that they have learned. In this session, you will channel students toward writing to teach others about their discoveries. You'll emphasize the importance of writing precise procedures so their experiments could be replicated. You'll introduce mentor texts for students not only so that they can see how real-world lab reports go but also so that they might revisit and improve lab reports already in progress.

Another reason we use mentor texts in this bend is to highlight the value of using charts, tables, and graphs. Organizing data visually makes relatively detailed information clear for readers. Even more important, it will help your students understand the results of their experiments—they will be able to tease out patterns, ask better questions, and analyze their results. Accordingly, you'll want to choose mentor texts that are not necessarily connected to your forces and motion topic but that include examples of the qualities of informational writing that can be transferred to your students' booklets. Be sure to select texts that reflect the amount of writing your highest writers can aspire to. They will most likely contain some of the examples of elaboration work you can share with your class. By the end of this bend, your writers will be designing and writing up their own experiments and also *revising* those experiments using what they have learned about how scientists work.

"In this session, you will channel students toward writing to teach others about their discoveries."

Ultimately, this bend nudges students out of the nest of shared inquiry, giving them options and opportunities to use what they know about writing and scientific method independently. By using the charts you have developed in Bend I and by trying out the organizational strategies they notice in mentor texts, your writers will spend more time revising and improving their writing than they did earlier in the unit. They will be working less in concert than they were in the first bend. While all of their lab reports might not yet look expert, in this bend it will be important to celebrate your students' discovery that rereading, rethinking, and revising are not just a writer's habits but a scientist's, too.

MINILESSON

Remember All You Know about Science and about Scientific Writing for New Experiments

CONNECTION

Remind students that they have "published" their results by sending them into the community, and rally their enthusiasm to do so again, with even more clarity, with another set of experiments.

"Writers, you have begun to get your scientific thinking into the world, haven't you! This morning as I was passing a third-grade class, I heard some of those students comparing the results they got from an experiment to the results you got in your experiments—by making your research public, you have now joined the scientific community of this school. Soon you'll be able to consider the results some of the other classes have gotten as they do your same experiments! I wonder what we'll discover then?

"For this next section of the unit, I want to invite you to publish another set of results—to get another scientific study underway in our community. You're more experienced now at experimenting and writing about your work, so this time you'll be on your own even more, and I'll teach you a little more about sharing your discoveries with readers.

"Remember on the very first day of this unit I told you that when scientists conduct experiments, they have a certain way they usually write to help them—they use a lab report format? They record what they expect to happen in an experiment, and they record what they actually do in the experiment, then they record how things go and what they learn." I pointed to our chart.

To Write Like a Scientist . . .

1. Ask a question about how the world works.
2. Record a hypothesis, a guess.
3. How will you test it? Record your procedure.
4. Conduct multiple trials, and record your results.
5. Analyze your results, and write a conclusion.

◆ COACHING

As much as possible, you want to bring your students into the life of scientific researchers, and certainly a big reason to make one's procedure public is so that others can conduct independent trials and their results can be compared. So this is a big deal.

Name the teaching point.

"Today I want to teach you that when scientists conduct an experiment, they remember all they know not only about science itself but about writing about science, too."

TEACHING AND ACTIVE ENGAGEMENT

Ask children to bring past knowledge and experience, both to hypothesize and to plan their writing about this experiment.

"So today, and forever more, you'll think back on what you know, what you've learned, and how it can make your work stronger. This will be your question for today: Using the catapults we build, which goes farther, the ping-pong ball or the cotton ball?

"So, we have our question, and we have two kinds of things to think about." I held up a hand on either side of me. "One thing we have to think about," I gestured with one hand, "is what do we know scientifically, about the question? What can we predict will happen, based on our previous experience and experiments? The other thing we have to think about," I gestured with my other hand, "is the writing part. What do we know about scientific writing that we can use to capture what we are doing and share it with the scientific community we are part of? Right now, turn to your partner and talk about both things—what is your hypothesis about what will happen, and how do you plan to write about it?" As they talked, I made sure our "To Write Like a Scientist" chart was visible.

I squatted down to listen to a partnership. Rebecca said, "We can test it out just like we tested the car on the carpet and on the floor!"

Jenna responded, "We can build the catapult, then we can try the cotton ball. We can use the meter stick to measure again, right? Then, we can write it."

Another partnership was looking at the chart. "We can write our hypothesis. I am going to write that I think the ping-pong ball will go farther because it is heavier." said Porter. Jacob agreed.

"Researchers, you are thinking back to your experiences and experiments and using that thinking to predict what will happen. And you are looking at our chart to remind yourselves of the scientific process to plan each page of the writing you will do. That is scientific thinking! I heard many of you saying aloud the question and hypothesis you will work with today. You'll have a chance to write that in a bit."

Channel children to plan and record a procedure for testing their hypothesis.

"The next thing that scientists tend to do after identifying a problem and hypothesizing is ask, 'How can we test this?' Scientists think about the procedure they will follow to test out the question—and that's hard.

Do not gasp at the thought of students launching objects across the room. Children will want to be included in this investigation. The level of engagement will continue to skyrocket, and your students' writing will soar.

Notice that this teaching echoes the teaching from the start of the unit. This is deliberate. The writing process and the research process and the learning process—they are all cyclical.

"In a minute, I'm going to ask you for help conducting this experiment," I said. "Right now we have to figure out what the experiment will be. How will we test out whether the catapult will send the cotton ball farther or the ping-pong ball farther? Start designing the experiment out loud, with your partner." I let children talk for a moment.

Organize a fishbowl, with four volunteers going through the experiment that the class has planned, while you coach and the class records.

Instead of gleaning suggestions from the whole class, I reconfigured the group, as in the first session of the unit. "Writers, can you position yourself on the perimeter of the carpet?" As the children moved so they were framing the meeting area, I asked for four student volunteers to help set up the catapult.

Whenever possible, use structures and methods you have used before so that the organizational challenges don't overwhelm everything else. The fact that this system for altering the way one sits and for using volunteers was done in a recent session makes it much more powerful.

Speaking to the rest of the class, I said, "You all know what goes on the third part of your lab report—the procedure, right? Just like last time, use story boxes to quickly draw and label what you think we should do with all this stuff to make the catapult," and I gestured to the ruler, masking tape, spoon, and rubber bands.

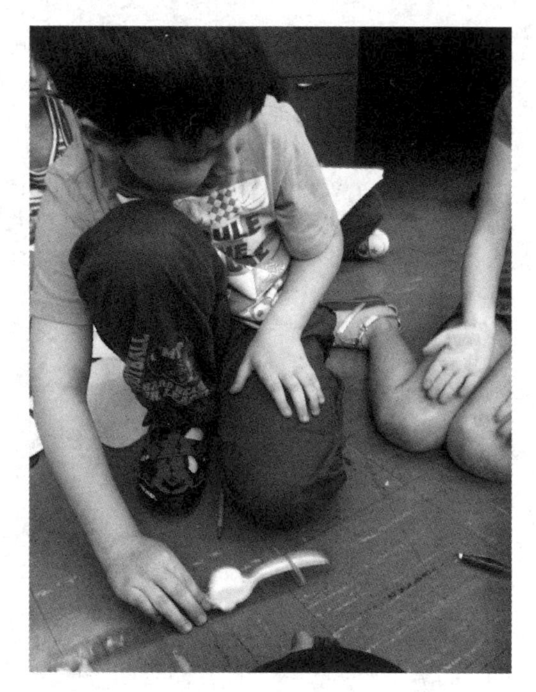

FIG. 7–1 The catapult experiment

The four volunteers sprang to the front of the rug area and got started while others wrote. Soon they'd laid the spoon on the ruler, wrapped the rubber band around them, taped the rubber band to the ruler, and placed the cotton ball in the spoon. They'd also pushed down on the end of the spoon and laid out the meter sticks, ready to measure the distance objects flew once they'd left the catapult.

Channel students to record their planned procedures, emphasizing the importance of precise procedures. Encourage them to record their results, including the unit of measurement.

"So class, let's listen to these four scientists' planned procedures and see if the rest of us agree with their plans." Turning to the foursome, I asked, "What will your procedures be?" One described how she'd built the catapult, and another added that first he'd fling cotton balls, then he'd fling ping-pong balls.

"So now you've heard what these kids plan to write, and you've written procedures before," I pointed to the chart.

In Procedures . . .

- Make a "You will need" section.
- Draw pictures that teach with labels, details.
- Number the steps.
- Include detailed measurements (2½ in.).
- Tell not only what to do, but how to do it.

"And you know how to write like a scientist, and if you forget, we've written the steps on our other anchor chart." I pointed to the chart.

LINK

Send children off to test their hypotheses, reminding them to write up their experiment so that others can use and replicate their results.

"Now I am going to give every partnership a baggie of equipment: a ruler, a cotton ball, masking tape, a ping-pong ball, a rubber band, and a plastic spoon. Your job is to work with your partners to make a catapult and test your hypothesis, all while jotting notes. And then you'll go back and fill in or actually write out the whole lab report. Ready?" The children nodded enthusiastically.

"Remember, this is important work, and you need to make sure that you are documenting each step of the process. You want to make sure that anyone can replicate it (that means do it again, exactly the same way), so get every bit down." I then gestured toward the "To Write Like a Scientist" chart and reread it to the students one final time. "I want you remember to use this chart to keep you on course. I suggest taking just a few minutes to work on your catapults. Then, you can spend the bulk of your time writing. Off you go!"

You may find your children have some difficulties measuring how far the ping-pong ball flies, since it will undoubtedly bounce or roll much farther than the length of the meter sticks. You can let children problem solve here! Or, you could suggest they measure where the ball first falls after launching—though that happens quickly.

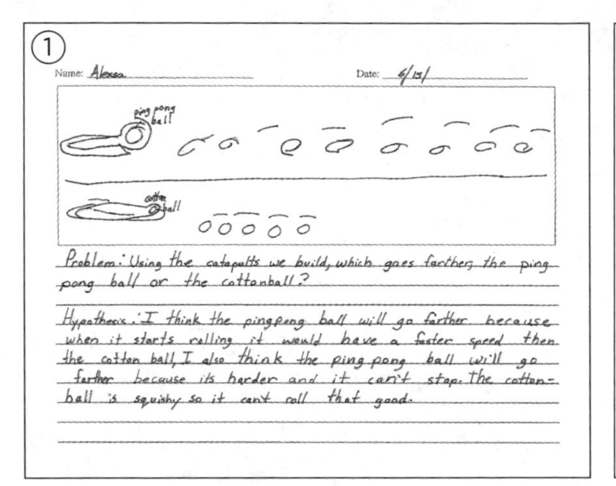

Problem: Using the catapults we build, which goes farther, the ping-pong ball or the cotton ball? Hypothesis: I think the ping-pong ball will go farther because when it starts rolling it would have a faster speed than the cotton ball, I also think the ping-pong ball will go farther because its harder and it can't stop. The cotton ball is squishy so it can't roll that good. LABELS: ping-pong ball, cotton ball.

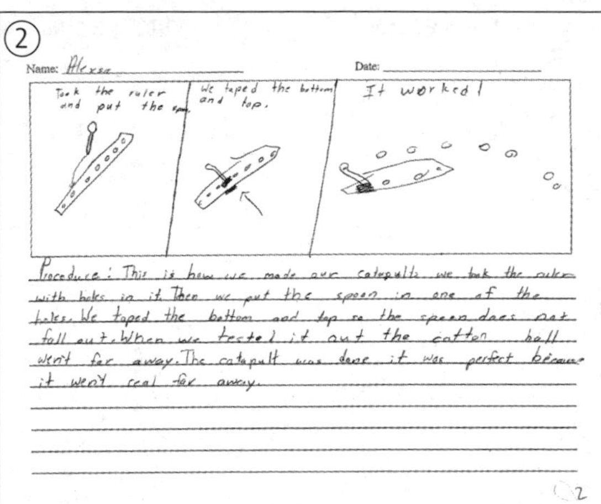

Procedure: This is how we made our catapults we took the ruler with holes in it. Then we put the spoon in one of the holes. We taped the bottom and top so the spoon does not fall out. When we tested it out the cotton ball went far away. The catapult was done it was perfect because it went real far away. LABELS: Took the ruler and put the spoon. We taped the bottom and top. It worked!

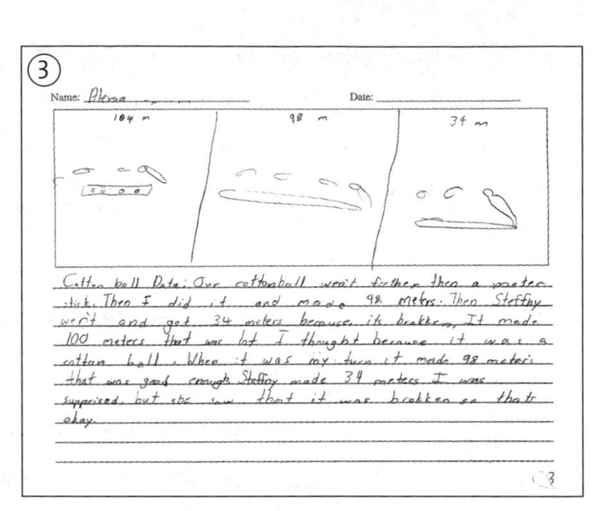

Cotton ball data: Our cotton ball went farther than a meter stick. Then I did it and made 98 meters. Then Steffny went and got 34 meters because it's broken. It made 100 meters that was lot I thought because it was a cotton ball. When it was my turn it made 98 meters that was good enough. Steffny made 34 meters. I was surprised, but she saw that it was broken so that's okay. LABELS: 104 meters, 98 meters, 34 meters

Ping-pong ball data: Everyone measured ping-pong balls everyone measured over 100 meters because it kept on rolling. When I did mine it went off to the side so it was hard to see that it was over a hundred. The last one the spoon broke. When I held the spoon and I watched it fly and it rolled and rolled and rolled. It went very far away.

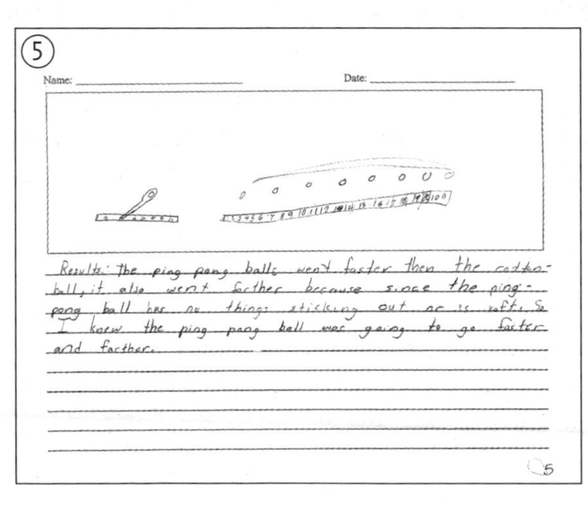

Results: The ping-pong balls went faster than the cotton balls It also went farther because since the ping-pong ball has no things sticking out or is soft. So I knew the ping-pong ball was going to go faster and farther.

Conclusion: My hypothesis is correct because I knew the ping pong ball was going to be faster than the cotton ball and I was right! The cotton ball on the first time was 103 m the second time was 98 and the third time was 34 m. We did the ping pong ball next. They were all over one hundred. We did the experiment 3 times each it was fun.

FIG. 7–2 Alexsa's lab report

CONFERRING AND SMALL-GROUP WORK

Steering Students' Attention to Data to Think and Write More in Conclusions

As YOUR STUDENTS have had the opportunity to conduct and write about multiple experiments in the first bend, most if not all of them will probably jump right into today's work without issue. As you scan the room and listen in on partnerships, you'll likely notice that students are moving through the scientific and writing process quite smoothly. Today, then, you may want to work with partnerships and individual students in bringing a closer eye to their lab reports to grow their scientific thinking and lift the level of their writing.

For example, you might come across several students who have recorded their results and are working on conclusions. In an earlier session, students had the opportunity to study a mentor conclusion, pulling out and trying on different elements, including reflecting on their hypotheses and providing multiple possible explanations for their results. Now that they've had time to practice these skills in previous lab reports, some of them may be ready to think more about the data they collected and write more about this in their conclusions.

I sat down at the rug beside Emily, who was writing her conclusions page and asked her to tell me how her conclusions were going so I could gauge what she was using from previous sessions. She replied, "I wrote that my hypothesis was right because the ping-pong ball flew farther than the cotton ball."

"It is pretty special for me to see that you didn't just notice what the ping-pong ball did, you thought about what that said about your hypothesis. That sort of thinking about how your results relate to your hypothesis is very much what professional scientists do." I responded. "What else are you planning to write in your conclusions?" I asked, guiding her to lay out a plan before she continued writing.

"I can write about how far the cotton ball went and how far the ping-pong ball went," she added.

"Okay! Where will you get that information?" I waited as she flipped to her results page. Emily recapped, "Well, our cotton ball went 60 cm the first time, 52 cm the second time, and 24 cm the third time. And our ping-pong ball only went 15 cm the first time, but then it went 268 cm the second time and 240 cm the third time."

MID-WORKSHOP TEACHING
Using Tables to Organize Information

"Writers, let me stop you for a moment. Jonathan and Jacob are having a disagreement about how to write up the results of their experiment. Jonathan wants to write about the cotton ball launch first, but Jacob thinks that the ping-pong ball results should be first, because the ping-pong ball went farther than the cotton ball. I'm wondering if you might look at their data and help them come up with a solution?"

I asked each boy to write the data from his half of the experiment where everyone could see it at the front of the room. The data looked like this: "210 279 259 64 80 79." As the boys wrote on the board, their classmates whispered, "Which one is which?"

"I'm glad you're asking questions, writers. Could you turn and talk with your partner about the data that Jonathan and Jacob found in their experiment? How could they organize this information to make it easier to understand? Remember, you want to be able to share your results with our scientific community." The room was abuzz.

After just a few moments, Michael raised his hand, "I think the best thing to do would be to write about the results together instead of just writing a few lines about each one."

(continues)

"I know something else," Gabriella said. "You really need to say which one is the ping-pong ball and which one is the cotton ball."

"Also, you need to put if it's in centimeters or not," said Alexsa.

I coached the boys to recopy their data in side-by-side columns, asking them to label their information by identifying the different balls and writing "First Try," "Second Try," and "Third Try" in the left-hand column.

	Ping-Pong Ball	**Cotton Ball**
First Try	210 cm	64 cm
Second Try	279 cm	80 cm
Third Try	259 cm	79 cm

Gabriella stopped writing suddenly, "The first try was the shortest one for both of them!"

"Wow, you noticed a pattern! If you label your information and put the data side-by-side in a table, that makes it a lot easier to see patterns. Seeing patterns is something that is useful to us all through life—in science and writing . . . and beyond! Writers, now and forever, remember that when you are recording data, you need to organize it carefully. Using a table will help you to keep track of your results so that you can understand the information you have found. It will also make it easier for your readers to understand your experiment."

Noticing the differences in some of her data, I responded, "Oh, wow! Well, there's a big difference in how far your ping-pong ball went in the first trial compared to how far it went in the second and third trials. Did you write about that in your conclusions?"

"Well . . . no," Emily responded. "I wrote that the ping-pong ball went farther than the cotton ball the second and third time and that I think that because it's made of plastic and flies better than the cotton ball."

"Okay. You are thinking about your results overall and comparing the cotton ball and ping-pong ball, which is great scientific work!" Noticing that Emily was capable of saying more, I extended, "I think you can push your thinking even further. When something in your data is completely different than the rest of your other data, this is called an anomaly. Scientists look closely at the anomalies in their data and ask, 'What happened?' to explore possible explanations to add in their conclusions."

I then coached her to think more about her results to produce more ideas to add to her writing. "What do you think happened? The result of your first ping-pong ball trial turned out so differently than the other two trials. Why do you think that is?"

After a moment, Emily said, "Well, maybe it's because it was the first trial, and I didn't have practice launching the catapult and didn't shoot it the best."

"Great! Get that down in your conclusions." I acknowledged her possible explanation and reminded her, "Now, you can think about other possible explanations, then you can look through your data again and notice any other anomalies and write about those, too!" I left her to finish writing on her own.

SHARE

Using Charts and Tables to Present Data

Offer students an example of adapting a chart or table to help analyze and interpret data. Suggest that they, too, can innovate to explain and present their data.

"Writers, you've accomplished a lot today—you are writing down the results of your experiments in tables so that you can keep track of the precise measurements you found. Take a moment now to share your results with another group. Notice how your results compare to others!"

After a few minutes, I interrupted their talk, "What I hear many of you saying, writers, is that the results for your ping-pong trials are very similar, and you are not surprised! In fact, as I walked around I noticed that a few of you have already written your conclusions. I would like for you all to look at the way Alexsa and Porter have presented their information. When you see what they have done, think about our 'In Conclusions' chart," I motioned to the chart on the wall, "and see if this helps you come up with some ideas for *your* conclusion."

Alexsa and Porter came to the front and displayed a table they used to analyze their data.

	Ping-Pong Ball	Cotton Ball	Difference
First Try	280 cm	46 cm	234 cm
Second Try	310 cm	72 cm	238 cm
Third Try	229 cm	79 cm	150 cm

"Alexsa and Porter made a guess that the ping-pong ball would always go farther than the cotton ball. Their guess was right! To write their conclusion they added a row to their results table showing how much farther the ping-pong ball went than the cotton ball. Now they can use these details in their writing.

"Remember, when nonfiction writers want to organize and understand the data they have collected, they often use charts or tables. Charts and tables can help you to present your information in a clear, precise way, but they can also help you answer questions you have about your results, the questions you asked at the beginning of your experiment."

Session 8

Studying a Mentor Text

The "Results" Page

Dear Teachers,

In place of a step-by-step transcript of the teaching that happened on this day, we give you this letter that aims to provide a vision for how your workshop may go. Our purpose for this is to support your independence, with hopes that you will then do the very same for your students. This session follows an inquiry structure (the one we used earlier in Sessions 2 and 4 and in the preceding *Lessons from the Masters: Improving Narrative Writing* unit). You may want to refer to those sessions to help you construct this one.

In the guided inquiries you led your students through in the last bend, you showed your writers how they can look to published authors when they want to know how a kind of writing goes. In this session, you will want to repeat a similar inquiry structure, this time giving your writers options for writing well off of their data and for presenting their results and findings.

MINILESSON

In the connection you will talk up the importance of organization, perhaps pointing out the various ways in which the classroom is organized so that children have a concrete example of systems they know well. You might say something like, "This morning I was looking around the room, and I began noticing all the different ways we have for organizing things—we organize our supplies in cubbies, we organize our folders in bins, we organize ourselves in meeting spots."

Then you will quickly shift your attention to scientific writing structures, so that children understand the focus of today's lesson. Perhaps you'll say, "That got me thinking about other ways we organize things, especially when we are writing. I started thinking about how scientists organize their writing. We explored this a little with the procedure page and the conclusion page. Remember how we noticed what writers did when they wrote those pages, and then we went back into our own writing and revised to make it

Common Core State Standards: W.2.2, W.2.8, W.3.4, RI.2.6, RI.2.7, RI.3.7, SL.2.1, SL.2.2, L.2.1, L.2.2, L.2.3, L.2.5

stronger?" Here you will be emphasizing the effectiveness of looking to mentors to improve writing. You might then affirm that improvement by asking students to quickly self-assess their own procedural pages using the "In Procedures" chart you created in Session 2. Tell children to give themselves a pat on the back to acknowledge all that they've learned to do to organize their writing. Be sure to keep the connection short, positive, and celebratory so that you have adequate time and energy to carry your whole class through the inquiry teaching for the day.

You may want to refer to the previous inquiry sessions, using those as models to draft an inquiry question for today. Be sure the wording is easy for your students to understand. Or you could take one of our suggestions: "How do procedures go?" or "How do scientists organize their results?" Whatever you decide, it is helpful to have recorded the question in advance and to reveal it as you read it aloud.

Say something like, "Today I brought in an example of a results page for you to study. I want you to study this results page really closely and then answer this question: How do scientists organize their results?"

During the teaching and active engagement, children will work in partnerships to study the mentor text, focusing on particular aspects. Set them up by pointing out some features you hope they will notice. Say, "Let's take a look at this mentor text. What do you notice about how this scientist organized her results? Look at the top half of the page. Hmm . . . it looks like there is a title to tell us what this is. Look at the visuals and read the words and numbers. What do you notice? Turn and talk with your partner!"

Children will probably notice that there is a table with columns and rows of numbers. They may realize that the numbers indicate the findings of the experiment and name what, exactly, these tell (for example, "The numbers say how many hours in each section on the table.") and perhaps the specific categories (for example, "Yeah, the hours for how much electricity people used on the weekends and weekdays."). Chances are your students will at some point connect what they notice on the mentor results page with the results pages for their own experiment. If not, you will want to steer them in this direction. At this point, they may notice ways in which their own result page can draw on similar features as the mentor one, as well as ways in which it will need to differ.

After a few minutes, suggest that children turn their attention to the rest of the results page: "Now look at the second half of the page. Look! There's a title . . . it's the same as the chart above. Hmm . . . what else do you notice? Turn and talk with your partner!"

Give children a few more minutes of partner talk before calling back the class to share out some observations. Children might notice that in addition to a table there is a bar graph or a chart (similar to the ones they've seen in math) or that there's a key that tells what the colors on the graph or chart represent. They will probably see the words on the side of this additional tool and realize that these are labels, and they should see the title at the top. If they aren't doing so on their own, nudge them to think and talk about what roles these various features play to the overall organization of the results—how they help to orient and instruct someone reading this page and hoping to learn or to duplicate the experiment. You might say, "Wow! You are noticing so much about the way this scientist organized her data. Hmm . . . I wonder why she organized her results page this way? Why do you think a scientist would organize information this way?"

By this point in the unit, children will probably say that haphazardly putting the information onto the page would make it confusing and messy, and that the title and labels explain what the numbers mean, and that the table/graph/chart clearly displays the results.

If your students have learned about charts and graphs during math or science instruction, they will have a ton of knowledge to bring to this lesson. You will want to draw on this, referring to the math or science units and to charts and graphs around the room. Push your students to think about why people organize information in these ways and again remind them to focus on the *organization* of the data rather than the data itself, which is sure to bring up questions and cause excitement around the numbers and visuals. To dispel this, you may also want to give children a brief explanation of the content and then shift their focus to the way the information is presented. Give them time to study the mentor text closely and to talk with their partners so that they grasp how the information is organized and how they might draw on their discoveries to organize their own results pages today.

For the link, suggest some choices that students have based on what they've discovered in their inquiry. Be sure that children have a repertoire of things to do as they go off to write. You might say, "Okay, writers! Think about what you might try today based on your observations from this mentor text. You can go back and revise your results pages for yesterday's experiment in a way that is clear and easy to read. You may even decide to go back to earlier lab reports and revise those results pages, too! Decide what you're going to work on first, then get started!"

CONFERRING AND SMALL-GROUP WORK

As you confer today, keep an eye on children's volume. Students who work slowly or who can easily get off track, not producing much writing, may need some prompting. Check in on these kids, and ask them what their goal is for the rest of writing workshop today. If a child replies, "One page," suggest that he might aim for even more the next day. Say, "Think you can beat that tomorrow? How about one and a half pages?" Remind these students that writers set goals to push themselves to produce more using all they know about writing.

Other children might need some help thinking between the mentor text results page and their own results page, to understand the ways in which the categories and features connect despite the difference in content. With some nudging, a child might realize that the amount of electricity charted for weekend vs. weekday use would translate to the distance traveled by the ping-pong ball in one column and the cotton ball in the other. That is, the same tool can be used to track different amounts or lengths. Likewise, in both charts, there might be a third column that shows the differences (between hours of electricity use in the one, and between centimeters traveled in the other).

MID-WORKSHOP TEACHING

For the mid-workshop teaching, you might remind children to use precise language as they write. In just a few days, you will teach a session on using domain-specific language, during which you will focus on the terms that are specific to a unit on forces and motion. For this session, you might simply say that just as scientists are accurate in their experiments, they are also precise about the words they use to describe things. They don't simply say that an object went a long distance. They name the object, tell how far it traveled, and use words to describe the way in which the object moved (for example, the ping-pong ball bounced down the ramp, then skidded across the bare floor, traveling sixty-two centimeters).

SHARE

During the share today, you might ask children to think a bit about how the tables, charts, and graphs in the world are organized and why. What does this teach? How can these tools help people to organize/understand things in the world? Elicit responses. If children come up empty-handed, encourage them to think about what they've noticed in the books they are reading, in the classroom, and in other subjects (math, science). Students will probably notice that these tools group information in ways that people can learn. They will probably point out that charts and graphs often include numbers and record amounts of things. Some record not numbers but information people want to remember, like names and birthdays. Celebrate whatever your children notice, and encourage them to be on the lookout for other kinds of tables, charts, and graphs not just in the classroom but at home and in the world.

Good luck!

Lucy, Lauren, and Monique

Session 9

Comparing Results and Reading More Expert Materials to Consider New Questions

IN THIS SESSION, you'll teach students that scientists compare the results of their experiments against other scientists' results, using these comparisons to grow and extend their thinking.

GETTING READY

- ✔ Your own lab report and a student's lab report to compare results (see Teaching)
- ✔ The results pages of both lab reports, projected on an overhead or copies of these pages to hand out to students
- ✔ Set up new writing partnerships for today so that students can compare the results of their experiments with a different set of results (see Link).
- ✔ Sources related to the science of the catapult (speed, gravity, angles), such as pages from books, websites, articles, along with captioned and labeled photographs to hand out in the mid-workshop teaching (in addition to several new texts, you may want to include some of those distributed in Session 5)
- ✔ Post-it notes for students to jot down new information gathered from the scientific sources (see Conferring)
- ✔ List of prompts to help children connect information from sources to their experiment conclusions (see Conferring)

Common Core State Standards: W.2.2, W.2.8, RI.2.1, RI.2.3, RI.2.9, SL.2.1, SL.2.3, SL.2.4, SL.3.4, L.2.1, L.2.2, L.2.3, L.2.6

TODAY'S SESSION BUILDS ON SESSION 3 in Bend I in that it sets up children to pose new questions as they consider new information. Students will learn that scientists don't work in isolation; they don't just conduct an experiment, record results, and then reflect solely on those. Rather, they extend their thinking by considering other sources side by side their own work, letting this new information spark new thinking.

In the minilesson, you will teach children that scientists study their own results against the results of other scientists who have conducted the same experiment. They compare both sets of data, noting what's different and then pushing themselves to consider possible explanations. Through this work, they come up with new questions, new thoughts, new possibilities. In effect, you will be teaching children the importance to scientific writing of analyzing and revising and thinking in the face of new information. This is cognitively sophisticated work—what Norman Webb describes as Level 4 in his Depth of Knowledge (DOK) framework.

The second half of the workshop extends this work. Children will put their writing on hold to read materials about forces and motion, taking notes on any new information they learn that relates to their catapult experiment or that raises new questions. Again, children will learn to think in new ways, deepening their understandings as they consider possibilities they may have missed earlier.

Rely especially on partnerships today so that children work together to grow their thinking. Encourage them to speak and listen in ways that build on each other's thinking, while simultaneously learning from outside experts. Expect that some students will need scaffolds, and be prepared to offer concrete tools such as prompts to help them extend their thinking. Meanwhile, also expect that today's work will energize students. There's nothing like the discovery of new questions, new hypotheses, and new explorations to enliven work that may be beginning to feel repetitive.

MINILESSON

Comparing Results and Reading More Expert Materials to Consider New Questions

CONNECTION

Remind children of the work they did in the previous session, and suggest that they are ready for the next step.

"In our last session you spent a lot of time working on your results by trying out things from a mentor text and organizing all your data into clear and easy-to-read tables and charts. Looking around the room, I saw lots of neat rows of numbers, of results. This room is full of scientific writing. And you know what? I think you're ready to do more than just record and think about your own data. I think you're ready to think *across* data. Scientists don't just do an experiment all by themselves and that's the end of it! You know already that scientists work within a community and share information with each other all the time."

Name the teaching point.

"Today I want to teach you that scientists compare their results with the results of other scientists who have done related experiments, asking, 'How do these results connect to my results?' and then they come up with new ideas to explore and new questions to answer."

TEACHING

Compare your results with those of a child in the class who conducted a related experiment, looking for connections and suggesting possible reasons why.

"I'm going to look at my results page and compare it with Esteban's results page from the experiment we did with the car and ramp. Watch me as I look at the results for each trial, both on the carpet and on the bare floor, and ask myself 'Why are my results different?'"

◆ COACHING

In this lesson, we aim to teach children the value of comparing results, thinking across data, and reading widely about the topic at hand. But perhaps even more importantly, we aim to teach children that scientists' work does not take place in a vacuum, in isolation. Scientists are part of a community that shares results, works together to solve problems, and reads widely to gather as much information as possible. We aim for students to feel like (and genuinely become) members of this community.

I showed both my results and Esteban's (see Figure 9–1) from the car and ramp experiment, pointing at each measurement to show students which numbers I was looking at.

Esteban's results

We tried this four times on the rug. The car traveled 13 cm, 14 cm, and $14\frac{1}{2}$ cm.

My results page: On the carpet:

First trial	57 cm
Second trial	62 cm
Third trial	51 cm

On the first trial, the car went 57 cm. On the second trial, it went 62 cm, and on the 3rd trial it went 51 cm.

"Hmm . . . My results say the car went 57 cm on the carpet, but Esteban's went 13 cm on the carpet. On the second trial, my car went 62 cm, but Esteban's went 14 cm! On the third trial, mine went 51 cm, *but* Esteban's went $14\frac{1}{2}$ cm.

"I wonder: why are my results different from Esteban's? My car went farther than his on all three trials on the carpet! Could it be that the carpet was smoother where I tested it? He used a really fuzzy rug. Could it be the fuzzy rug slowed his car down more? I'm going to write these comparisons and questions in my conclusion page.

"This also gives me an idea for a new experiment—I could test out different patches of the carpet and see if the car goes different distances on different amounts of fuzziness! I'm going to put that under 'Future Investigations' in my conclusion."

Debrief—recall what you did in ways of thinking and writing students can duplicate in their conclusions.

"Writers, notice how I compared my results from the carpet trials to Esteban's results from the carpet trials. I looked at the results and asked, 'Why are mine different?' Then I made sure to record my comparison and questions on my conclusion page. And I added ideas to test out my 'Future Investigations' section of my conclusion."

ACTIVE ENGAGEMENT

Set children up to examine the next set of results you and the student each got, comparing, posing questions, and generating possible explanations.

"Let's keep going. Look at results from both Esteban's and my bare floor trials. Work with your partner to compare our results, ask questions, and think of possible explanations—and even ideas to test out our explanations." After a few minutes, I asked students to share their conversations.

FIG. 9–1 Esteban's results

Stella said, "On the third trial your car went 280 cm, but Esteban's car only went 246 cm." Henry chimed in that on the second trial, my car had gone 267 cm while Esteban's car had gone only 210 cm. Magali pointed out that my car had gone farther in most of the trials on the bare floor.

"You've noticed a pattern, Magali. Hmmm . . . what are you all wondering, and how could you test it?" I said and let my voice trail off to give everyone a moment to think. Jonathan's arm was waving madly. I indicated that he should share.

"Why did your car go farther on the bare floor?"

"I'm wondering that, too. Do any of you have thoughts? What might explain the difference in the distance my car traveled and the distance Esteban's car traveled on the bare floor?"

There was a moment of silence before Nina volunteered, "Maybe they had different-size cars?"

"What a thought, Nina! Everyone, think for a moment about how we could test that and how we could write about that in our conclusions and 'Further Investigations' sections." I paused while we all thought, silently.

It is important not to be afraid of a little silence. We are asking students to do deep thinking work, and that can take time.

"Yeah," Henry said, "Or maybe the bare floor was more slippery when you tested the car?"

"Interesting, Henry! What would we write in conclusions and 'Future Investigations' to go with that idea?"

We continued in this way until share time was over.

LINK

Remind children of the work they have done today, and pair them up in new partnerships so that they can try out the work of comparing results with a different "scientist." Suggest several paths children might take as they do this work.

"Writers, you are thinking like scientists. Remember that one thing you can do during an experiment is to compare results to learn more and also come up with new questions to explore and answer. Just for today, I'm going to pair you up with someone from a different partnership so you can compare results with that person. It can help push your thinking to have a new scientist partner. Once you have your partner for today, have a go at this work of comparing results and adding comparisons, questions, and possible explanations to your results and conclusion pages. If you finish comparing results from one experiment, you can go back and compare results for other experiments.

"Some of you may find that you end up revising your ideas and understandings through talking with a new partner. Others of you may come up with new lines of questions. It will be fun to see what comes of this new pairing! Remember to record your questions, comparisons, and possible explanations in a clear way so that anyone reading your conclusion can understand your thinking. Off you go!"

CONFERRING AND SMALL-GROUP WORK

Helping Children Use Information from Sources to Strengthen Their Conclusions

IN TODAY'S MINILESSON you asked children to consider the results from the same experiment of two different people in the room (you and a child), to think about how these differ, and to then generate new questions and ideas to explore based on the data. You've suggested that this work—of comparing results—is one way that scientists grow their thinking. Inherent in your teaching is the suggestion that scientists, like writers, revise their work based on new discoveries—a message that you will support when you set up students during the second half of the workshop to read materials on forces and motion and to then push themselves to revise their thinking, questioning, and ideas based on that new information. You'll want to carry over this spirit of revision to your conferences today.

As you move around the room, you'll want to especially be sure that children are carrying over the information they are learning in ways that in fact push their writing in new directions. This is challenging work for anyone, let alone seven-year-olds. You may want to have a few concrete tools on hand, as well as suggestions of how children can take notes in ways that set themselves up to make meaningful connections, ask thoughtful questions, and extend their thinking. Today, children are likely to need coaching like the kind called for in this small-group conference:

I pulled up to a table of students poring over each other's results as well as some articles, and filling Post-it notes with the information they learned. "Wow, Times Square Table! I see that you are furiously researching and taking notes about forces and motion. Look at all those notes! I think you're ready to start adding all that information you've learned to your writing. Ready to do that?" The kids nodded.

"I can Post-it the note on the page to show that it goes there," Henry said, demonstrating with one.

Jonathan said, "I'm going to write it at the bottom of the results page because this note about gravity goes with the results." Magali chimed in, a bit uncertainly, that she could add her notes to her conclusion.

MID-WORKSHOP TEACHING **Reading and Research**

"Writers, for the second half of today's workshop, we are going to try something new. Instead of *writing*, you are going to *read*. I'm going to give each table some materials on forces and motions that I gathered from articles, books, and websites. Scientists do this all the time. They gather new information by reading articles or short books or by watching videos related to the topic they are studying. They take in new information and from it they come up with new questions and thoughts. From these, they develop further study questions and then they conduct further experimentation—which leads to further revision to results and conclusions!"

As I placed a selection of materials on each table, I said, "As you read through these, take notes on any new information you learn, especially information that gets you thinking about your experiments in new ways. Put any information that pushes your thinking about your results—from any of your experiments!—onto Post-it notes. Be sure to pull out information that pushes you to think in new ways or that helps you explain what happened in scientific terms."

After a few minutes, I called out, "Writers, talk with your partner about how the information you're gathering relates to your experiment and what further questions it brings up for you. You might also be considering new possible explanations for what you observed. Talk about that, too."

"All of these ideas would certainly work! But I think you can push yourselves *even more* to make sure the information you've collected both relates to your experiment and makes your writing stronger. One way you can do that is to not just add the notes to a part, but to also say more about how the information *connects* or *explains* parts of your experiment. You can use these prompts to help you connect the information to what you have already written." I shared a list of prompts with the children:

- From the article/book I read, I learned . . . This connects to my experiment because . . .
- This explains . . .
- This helps to understand that . . .

"Let me show you a place where I did this in my writing, and then you can try it out in yours." I showed them my writing. "On my conclusion page, I added information to explain why I think my hypothesis turned out to be wrong. I wrote:

> My hypothesis was wrong. The cotton ball did not go farther than the ping-pong ball when I launched it from the catapult. Maybe this is because the cotton ball is fluffy and the ping-pong ball is plastic? FROM THE ARTICLE I READ, I learned that some things have more resistance when they are in the air because of the material or how the object is shaped. This CONNECTS to my experiment because the cotton ball and ping-pong ball are made of different materials, so I think that maybe plastic has less resistance in the air than cotton.

"Do you see how I added the new information I learned in a way that connects, using the prompts to help me? In writing about it in this way, I realized that I could push myself to say more so that the notes I'd taken from my reading connected to and strengthened my writing.

"Go ahead and try this out with the notes you took." I coached in as the students did this work. "Use the prompts!" and "How does that information connect to the experiment?" I left the group to keep going and moved on to another.

FIG. 9–2 Jonathan says more about his results by drawing on information he gleans from outside sources.

SHARE

Rehearsing Plans

As rehearsal and to generate excitement, ask writers to share their plans for new experiments.

"Writers, now that you have all this new information about forces and motion, *and* you have new knowledge of how to conduct experiments and write about your findings, I bet your heads are spinning. Thumbs up if you have even *more* questions and ideas you want to explore." Lots of thumbs went up.

"You've been working like scientists today, rushing to analyze your results, research new information, and add notes to your lab reports. Let's take a second to pause. Right now, think for a minute about a question you have and come up with a plan for how you might test that question to find an answer."

After a minute, I said, "Now turn to your partner and share your thinking. How would you do an experiement using everything you know about science and writing about science? Use detailed and precise procedures!"

After a few minutes, I elicited some responses from the class. "When I point my imaginary conductor's baton at you, share what you just told your partner. Remember to be really specific so all of us can imagine what you're describing."

I pointed to Henry who said, "I wondered what would happen if you launch the catapult in a more open space outside. Like, we could make the catapult again, but launch it in the schoolyard. We could stand at the wall and then launch it to see how far it would go outside, not inside the classroom."

I pointed to Magali. "We should change how we launch the catapult and measure how far the cotton ball would go. We could launch it sideways, then up, or like a little less sideways, or backward!"

Luca was up on his knees. I pointed to him and he said, "Or see how high the different kinds of balls would go! Like the cotton ball and ping-pong ball *and* golf ball *and* bouncy ball!"

I pointed to Jonathan. "What if we launched the cotton ball from the stairs? I think if we launch it higher, the cotton ball would go farther.

"You have all let your research and experiments lead you to new questions and experiments you could try! That's what scientists do, always thinking of ways to answer questions, solve problems, and design new experiments! Tomorrow, I am going to give you a chance to try out new experiments with catapults, just like we did with the cars and ramps."

Session 10

Designing and Writing a New Experiment

IN THE FIRST BEND of this unit, you encouraged students to think, and then rethink, not only their writing but their beliefs and thoughts about forces and motion. It is with this sort of work that scientists become experts in their field. They conduct multiple trials and consult with fellow experts. They refer to scientific videos, texts, and journals to deepen their knowledge. They reread their results and come up with theories to explain the phenomena they have studied. Scientists build on their earliest experiments, designing new ones to solidify or increase their understandings.

This session is built on the principles of thoughtful, sustained research. To develop expertise, a study has to go deep (and beyond one day)! So today, you will return to the catapult, traveling the familiar path of experimentation with motion, force, and gravity. On today's journey, however, your students will be more than just learners. They will be burgeoning experts of the catapult and will be on the lookout for patterns of motion as they go.

This session begins with a reflection on yesterday's catapult session. Returning to the catapult will help students develop expertise by affording them the opportunity to spend extended time on one specific model that they will try to make better. You will ask students to design a catapult that can propel an object farther than the previous models.

As your students work in partnerships to rise to this challenge, expect to be amazed by the thoughtful ideas they propose to their partners about *why* to design their catapult in a certain way or *why* to try a certain angle or *why* to work with a particular kind of material. Take advantage of these ideas! Nudge your students to explain aloud and say more about their hunches and decisions.

Your hope is that at the end of today's session, your scientists will be able to explain their results, developing conclusions that help them see patterns across days of this study. With so much data flying around your classroom, you'll want to capitalize on the reasoning that will deepen everyone's expertise—perhaps including your own!

IN THIS SESSION, you'll teach students that scientists revisit their initial experiments and ask, "What do I still wonder?" Then, they use their initial results and writing to generate new experiments.

GETTING READY

- ✓ Writing center filled with pens or sharpened pencils; stacks of blank five-page booklets; revision strips, revision flaps, or Post-it notes; and a separate tray of individual sheets of lined paper in case writers need to add pages
- ✓ Identify spots for students to set up, design, and test their catapult models; consider carpet space or cushioned landing surfaces.
- ✓ A baggie of supplies for the experiment: ruler, plastic spoon, rubber band, masking tape, ping-pong ball, and cotton ball—placed at each partnership's spot
- ✓ Clipboards for each student to easily jot down results throughout their catapult trials—especially for those not near a table
- ✓ Several meter or yardsticks, ideally one for each partnership, to measure the distances the objects fly
- ✓ A dry erase board and marker to write today's scientific problem: "How can we revise our catapult to make it shoot both the cotton ball and ping-pong ball farther than it went with the original design?" (see Teaching and Active Engagement)
- ✓ "To Write Like a Scientist" anchor chart (see Link)

Common Core State Standards: W.2.2, W.2.7, W.3.5, W.3.7, W.3.10, RI.2.1, RI.2.5, RI.2.7, SL.2.1, SL.2.4, SL.3.1, SL.3.4, L.2.1, L.2.2, L.2.3, L.2.6, L.3.6

MINILESSON

Designing and Writing a New Experiment

CONNECTION

Situate children in the work of the unit so far, and let them know that they can continue with their plans today.

"Writers, as you have been working with your catapults over the last few days, many of you have come up to me to tell me your ideas about how to make the catapult shoot farther. Some of you were thinking about materials you could change, and others were thinking about catapult designs you could change. You are living like scientists!

"Scientists are always thinking about their experiments, asking questions and designing new experiments. As you conducted multiple trials and experiments, you not only learned about science, you also learned about scientific writing. Did you know that scientists don't just revise their writing, they also revise their experiments? Today you will have the chance to think about your catapult design. Do you think you can revise the design of your catapult to make the projectile fly farther across the room? Some of you have been asking me if we could have a competition. Well, today is your lucky day. Are you ready for the challenge?"

Name the teaching point.

"Today, I want to teach you that scientists study their results to learn, think, write, and experiment more. They do this by first revisiting their experiment and asking, 'What am I wondering? What else do I want to find out? What is my plan?' Then, they experiment again."

◆ COACHING

It may be the case that your students have already begun innovating on their catapult design—which is a natural occurrence as children play—even if this "play" is in the service of a particular experiment with particular goals.

TEACHING AND ACTIVE ENGAGEMENT

Set writers up to explore a new problem.

"Now, writers, before I send you off to work with your partnerships, I want you to think about the problem you are going to solve." Then, I wrote the problem on the dry erase board:

> How can we revise our catapult to make it shoot both the cotton ball and ping-pong ball farther than they went with the original design?

"I know many of you have ideas about how to do this. Take a second to think about the catapult that you used. What could you change to make your catapult fling the cotton ball and ping-pong ball farther? What have you learned from your research that gives you an idea? What have you learned from comparing results with other people that gives you an idea? Think for a second, and when you feel like you have an idea, please put your thumb up so I know you are ready to talk to your partner."

Ask partners to say aloud the procedure for their revised experiment, discussing a variable they will change.

As soon as I noticed that most of the thumbs were raised, I said, "Turn and tell your partner what you've learned that gives you an idea for how to make the catapult shoot farther than it did with the original design and explain the procedure you will use to test that idea."

If you discover that your students could use more support with revising and innovating on their procedures, you may decide to ask volunteers to share their ideas with the whole class.

Henry said, "I want to put the plastic spoon in the hole of the ruler. This way, the spoon will be stuck and it will hold it in place and spring back. That would work better because the spoon is bendier than the rubber band."

Then I moved to another partnership and listened to Porter, "I think if we keep the catapult the same but we stick it to the ground, it will shoot farther. When it moves it wastes energy. We could even tape the ruler part of the catapult up on the back of the chair so that it doesn't move."

LINK

Remind students of the ways scientists structure their writing.

"Remember all you know about science and all you know about writing about science as you experiment and write today and from now on!" Then, I gestured and reread the "To Write Like a Scientist" chart.

To Write Like a Scientist . . .

1. Ask a question about how the world works.
2. Record a hypothesis, a guess.
3. How will you test it? Record your procedure.
4. Conduct multiple trials, and record your results.
5. Analyze your results, and write a conclusion.

Set writers up to design, conduct, and write up new experiments.

"Writers, today you will have some time to conduct new experiments. Then, you will have time to write up your experiment with your new catapults. Afterward, we can compare results at the end of the workshop and see which design was able to send the projectiles the farthest! Remember to keep in mind all you already know about scientific writing as you go." The students nearly bounced out of their spots. "Okay, are you ready? Get set and go!"

CONFERRING AND SMALL-GROUP WORK

Reminding Writers to Plan

As YOUR STUDENTS GO OFF TODAY, they will be excited both by the opportunity to design their own catapult and the challenge to create a catapult that carries loads the farthest. You have stressed that they think like scientists—analyzing previous experiments, pinpointing variables to change, and planning for a new set of trials. They may, in fact, be so excited about the new catapult experiment that they forget to write precise notes. As you move around the room, be ready to remind your writers that they must plan for their writing, too. A conference might go something like this one:

I stopped at Magali and Luca's table to see what kind of progress they were making; they were launching projectiles left and right. I intervened quickly. "Wow, how exciting that you've been able to dive right in! Before you start your trials, though, let's take a step back, rewind, and think about what you need to do. It's clear you planned how to test the catapult. Another thing scientists and writers do is plan how their writing will go, *before* they begin their trials, so they know how they will keep track of the information." I asked them if they could come up with a plan for their writing—what did they need to think about before they started their trials?

Magali explained that they needed to set up a table: "We need to have two sections—one for cotton ball and one for ping-pong ball." "And numbers 1, 2, 3 because we launch it three times," Luca added.

I pushed them to be even more specific, "Great! What else?"

Magali responded, "Oh! We need to write *ping-pong ball* and *cotton ball* so you know that side is the cotton ball and that side is the ping-pong ball." I waited as they drew their tables and made their plans.

"Do you see how slowing down to make a plan first can help keep you on track? Don't forget to do that whenever you conduct an experiment or take on an important project or piece of writing!"

MID-WORKSHOP TEACHING **Using Labels and Titles to Highlight Important Information, Including Failures**

"Writers, eyes up here. As I walk around I am noticing how much writing so many of you are adding to each page! Wow—so much! Can I tell you about some of the strategies writers in this classroom are using to write like real scientists? Just now, Magali was working on a diagram of her catapult. She labeled all of the parts of the diagram—the ruler, the spoon, the rubber band. This is really important! By using labels, Magali was being detailed and precise in all parts of her writing.

"But Magali did something else with her diagram that immediately caught my attention. Magali's first catapult didn't work, so she and her partner had to revise their design. I knew this new catapult was important to their experiment because Magali used a title at the top of her diagram that said 'New Improved Catapult.' She wrote her title in bold capital letters and underlined it in red, so that made me notice it! And she also made a labeled diagram of the catapult that didn't work. That was smart, too! Now we can all learn from these mistakes!

"Writing like a scientist means writing with precision in all parts of your writing, like using labels and titles. But don't forget, writers—you should also be using everything you know about how to emphasize important ideas and information to your reader, like using bold letters and colors. And explaining the things you've tried that *don't* work as well as the things that *do*!"

SHARE

Comparing Results

Divide the class in half, and ask each half to determine from their results whose catapult flung cotton balls the farthest.

"Scientists, you have had time for multiple trials and lots of recording of information—now we are ready to figure out which of all your different catapults was able to make the cotton ball fly the farthest distance! I'm going to divide you into two groups, right here in the meeting area. Will you go into your groups and talk over your data to figure out which of the catapults in your group worked the best? Which consistently flung the cotton balls the farthest? In a moment, be ready to share your answer." I crouched down, listened, and watched while the groups talked and swapped lab reports.

Ask volunteers to recreate the experiments of the best catapult from each group to see if the winning results can be repeated—and see if the lab reports can be followed!

"Okay, each group has their top contender—now we'll need to recreate these two experiments to see if we get the same results.

"Could the partnerships who received these results please hand us your lab report and catapult? (I know you haven't finished the conclusion part yet.) Now we need a different partnership to step up and recreate this experiment, using the lab report, to see if we get the same results. We'll just use your catapult rather than rebuilding it, for now. We'll also need someone to measure how far the cotton balls go and someone to record the results on the board for us." Hands shot up.

"And we need to do the same thing with this second catapult." After three trials each, with the results recorded, I said, "Let's take a look at the results. Wow! Luca and Magali's cotton ball went 143 cm—very far! Nina and Stella's cotton ball went 242 cm—even farther!"

Ask students to write down their ideas and hypotheses about why these two catapults shot farther than all the others and to connect these ideas to their own experiments.

"I have so many thoughts about why these two went so far, compared to the others, don't you? Take these last few moments to write some notes to yourself about what you are thinking and hypothesizing about your own experiment because of what you've just learned here. What can you take from this and use for your own conclusions? How can you make this connect to your own experiment?"

1 Nina

If I jast use the spoon as a catapult will the balls go farther? My hypotbest is it will go farther. I think so because a spoon has more presher because its plastic

2 you need spoon/plastic ping pong ball coton ball

lean the spoon with loaded coton ball shoot and shoot the ping pong ball.

3 ping pong ball cotton ball

Coton	Ping Pong
120 cm	132 cm
285 cm	500 cm
197 cm	284 cm

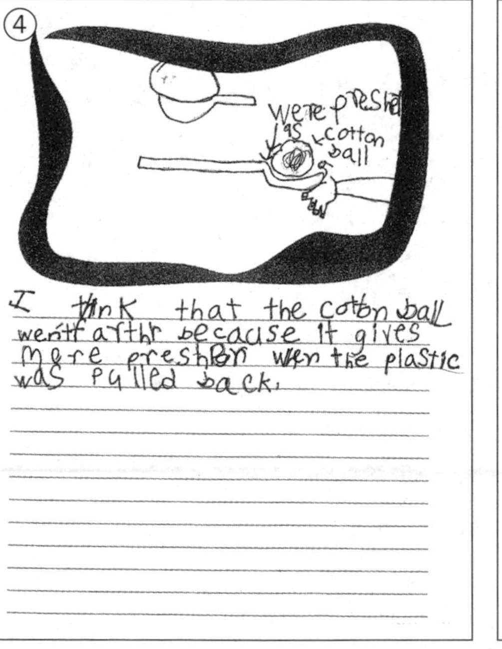

FIG. 10–1 Nina's catapult design—changes a variable

Session 11

Editing

Domain-Specific Language

ON THIS FINAL DAY OF BEND II, you'll want to celebrate all that children have tackled over the past week. Acknowledge that they are thinking and acting more and more like scientists each day. Then tell them that there is yet another thing they can do to assume this role; they can *speak* and *write* like scientists. Chances are, some or even many of your students are already incorporating technical terms they've learned over the course of this unit into both their speech and their lab reports. Today you'll shine a light on this, suggesting that scientists the world over share a particular lingo—and that the lingo often varies from field to field. As forces and motion experts, your children can now "talk the talk" of forces and motion.

There is nothing quite like incorporating precise, technical vocabulary into speech and writing to make one feel expert in a subject. Children will also enjoy sharing their own specialized language and knowing they can use it to teach others. You will want to capitalize on this, asking students to brainstorm and compile a list of words that they can use in their communications as scientists—in partner talk, in lab reports, and to teach others outside the field. Then, of course, you'll need to give children opportunities and invitations to use these words all across the day and even at home. As children talk and write, incorporating forces and motion lingo, they will naturally make mistakes; that's okay. All of us misuse words as we are first using them. Welcome your students' attempts and gently correct these as you notice them. Above all, have fun today!

IN THIS SESSION, you'll teach students that scientists use domain-specific language when speaking and writing about their topics. They do this so that they are as precise as they can be when talking about their experiments and to show their audience that they are experts in their field.

GETTING READY

- ✔ A dry erase easel or whiteboard with a marker to collect a list of relevant domain-specific words (see Active Engagement)
- ✔ A word bank that lists definitions for the collection of domain-specific words the class will generate during the active engagement
- ✔ A few information book mentor texts for students who may wish to study ways published authors introduce and teach specialized vocabulary (see Conferring and Small-Group Work)
- ✔ Writing center stocked revision tools such as pens, strips, flaps, single sheets of paper, Post-it notes, scissors, tape, and staplers (see Conferring and Small-Group Work)

COMMON CORE STATE STANDARDS: W.2.2, W.2.5, W.3.2.b, RI.2.4, SL.2.1, L.2.1.e, L.2.2, L.2.3, L.2.4.e, L.2.5, L.2.6, L.3.6

MINILESSON

Editing

Domain-Specific Language

CONNECTION

Liken the particular ways in which children talk about things they know well to how scientists talk about the subjects they study using specialized words.

"Writers, I've noticed something interesting. Sometimes in the morning when you talk to one another about television shows you watch or about computer games you play, you toss around terms lots of people might not understand, like 'Angry Birds,' 'Beyblades' and 'a ripper and rip cord.' It's almost as if you're speaking another language, a language specific to those games or shows! Well, that got me thinking. It's a lot like that for scientists, too. Scientists know and use all sorts of words that are specific to the kind of science they do. That's because those words are specialized ones that describe measures, activities, substances, parts—anything particular to a kind of science."

Name the teaching point.

"Today I want to teach you that scientists use expert words—called *technical vocabulary*—to make their writing and their teaching more precise. All of you, as forces and motion experts, can do this, too. You can begin to use words that are particular to the topic you are studying in both your discussions about that topic and in your writing about it. You can 'talk the talk.'"

TEACHING

Teach the concept of technical language, inviting children to brainstorm domain-specific terms they know on topics they know well.

"Writers, any subject matter has its own set of technical terms. For example, teachers have lots of our own technical terms. We assess your writing using *rubrics*. We call the class together for *minilessons*. We ask you to *turn and talk*, and when we do that, we mean something very specific to teaching. These aren't words and phrases that just anyone uses. These are terms that teachers use to talk about our subject.

"Right now, think of a subject you know well. This might be a hobby or a sport or a game you play. It might be a family tradition. Think of one subject that is an expertise for you. Give me a thumbs up when you've got something in mind.

◆ COACHING

Children will get a kick out of you throwing around terms from their world, and meanwhile—as is your goal—they'll have a clear model of what you mean by "technical vocabulary."

"Now turn and tell your partner at least five specialized words you know because of your expertise on that topic."

I listened in and heard Rebecca list dancing terms to Tyler, "cha cha, samba, jumpy jive, rumba . . . "

Then Tyler took a turn telling Rebecca some ice hockey terms: "puck, neck guard, rink . . . "

After a minute, I said, "Now see if you can come up with a sentence or two about your topics that include as many specialized words as you can think of—so many that those of us who don't know about your topic won't know what you are saying!

"Once you have a sentence or two in your head, try it out on your partner. See if he or she has any idea what you're talking about."

I asked Tyler to share his sentence. He said, "A *puck* is a round disk that hockey players slam their sticks against. Pucks move across the ice, and players try to slam it into the goal."

As I scanned the class, many other children had their thumbs up signaling they wanted to share. Rebecca volunteered, "The *cha-cha* is a dance you do with a partner and is a Latin-style ballroom dance."

"Wow, I didn't know any of this! See how each one of you has a specialized language you speak? When you know a lot about a topic, you use words that are foreign to people who are not 'in the know.' Everyone in this room also has a shared language about a topic that's become an expertise: forces and motion."

ACTIVE ENGAGEMENT

Redirect children's attention to the shared class topic, forces and motion, and together, generate a list of relevant domain-specific words.

"Right now, let's see if we can come up with a list of specialized words—technical vocabulary—that we know well as forces and motion experts. These will be words that we can use both when we talk and write about our catapult experiments. I know that some of you have already started including these kinds of words and phrases. I'll get us started." On the whiteboard I wrote, "Forces and Motion Lingo" and underneath it wrote, "push/pull." Then I opened up the work to the class. Soon, the children and I had generated this list: push/pull, friction, surface, pressure, force, gravity, motion, and balance.

"Okay, scientists, now it's time to 'talk the talk.' Let's try something fun. Turn and talk to your partner about PE class today, and as you do, try to use our forces and motion vocabulary words. You may even come up with new words that we can add to our list. Listen for those. Okay, turn and 'talk the talk!'"

As children talked, I listened in and overheard this conversation:

Students will enjoy generating words, though you may find that you need to offer a bit of support to focus their brainstorming on words specific to force and motion. The beauty of this exercise is that it sets up children to engage in higher-level depth of knowledge work, as they must transfer concepts they have learned in one domain to an outside context.

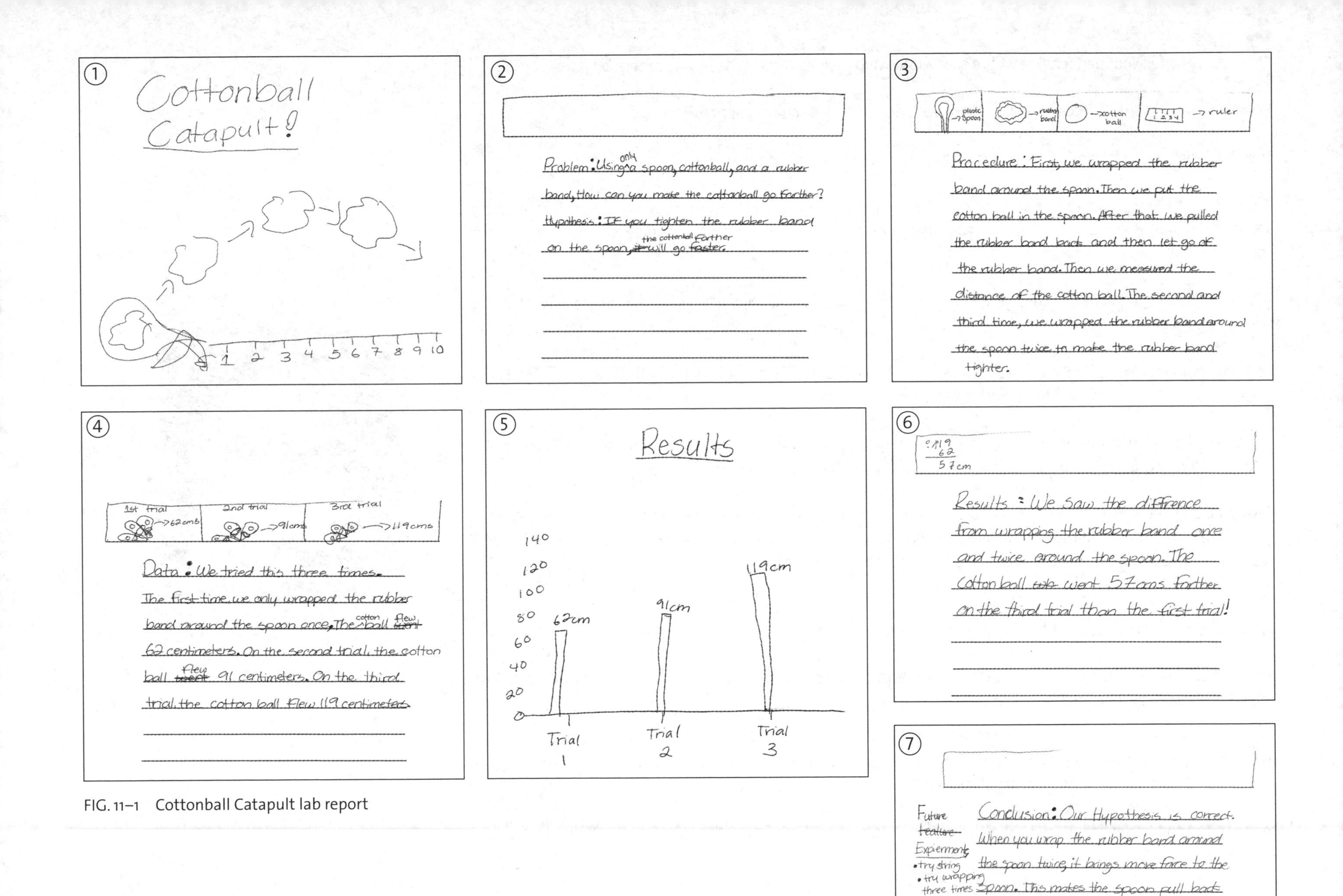

FIG. 11–1 Cottonball Catapult lab report

"When Matthew threw the ball at me, it came at me with a lot of *force.*"

"Yeah. Matthew is strong. I tried to get the ball from him, and we ending up doing a lot of *pushing* and *pulling* before I got it."

"Trayvon is strong too. He almost knocked me off *balance.*"

After a few minutes, I said, "Writers, you've got the hang of it. The forces and motion lingo is flowing in this room!"

Suggest that the class come up with a system for recording technical language.

"Now, if we're going to start gathering technical terms and using them not just to talk about PE class but to talk and write in precise, smart ways about scientific experiments, we'll need a system for recording the terms. It won't be enough to just put them in a list, either. I think we need a way to remember what these words mean, because some of them are new to us and kind of hard!

Hopefully many of our students will have a good idea of what they'll write about before today's lesson begins, and you'll be able to move them rather quickly into the minilesson in order to give them as much writing time as possible.

"I know! Let's make a word bank of forces and motion terms. Next to each word will go either an image that shows what it means or a simple definition. I'm going to work on that a bit while the rest of you write today."

LINK

Suggest that children review their work to be sure it includes forces and motion lingo—and if not, to incorporate it in clear, thoughtful ways.

"Today as you write, you might decide to keep an eye out for the words in our word bank. You might even notice more words or phrases we could add. If you do, tell me, and I'll put them in. If you haven't included any in your writing, see if you can add some in. Try to be as clear and precise as you can be so that anyone *not* in the know would be able to figure out some of these unfamiliar words by reading the sentences in which they appear. Some of you may want to go back and revise earlier sections of your report or old lab reports you wrote awhile back. Remember, you are now forces and motion experts, so use the lingo!"

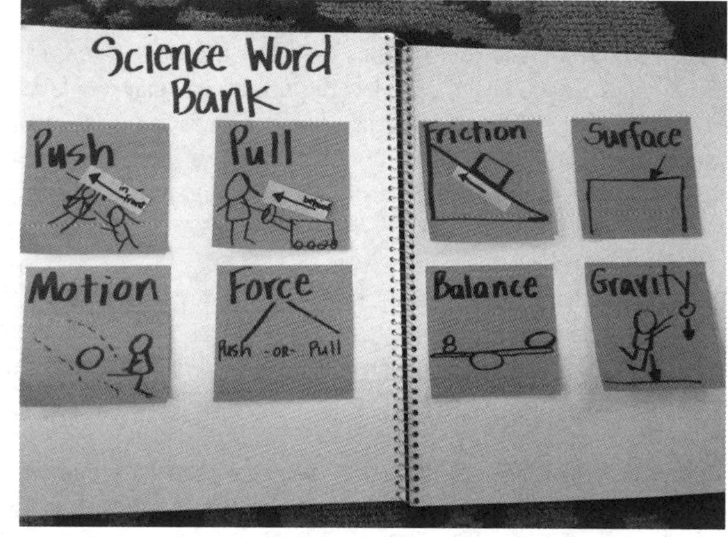

FIG. 11–2 Science word bank

CONFERRING AND SMALL-GROUP WORK

Supporting Writers as They Revise in a Variety of Ways

As YOUR STUDENTS GO OFF to independent writing time today, you may decide to invite students who feel like they need additional support to stay back and begin writing in the meeting area. This will enable your students to assess whether they need additional support. You may say something like, "Boys and girls, if you feel like you need a bit more help figuring out how to revise your writing to use the scientific expert words, you can stay back on the carpet and begin to work here."

This way, both you and your students will know the focus of today's conferences or small groups. As you begin to work with students, you may decide to teach them to reread bit by bit looking at the science word wall, asking themselves, "What is the science here? What word could I use?" Then, you can teach children to think about ways to insert this language. Students may decide to include not only the scientific word but also its definition.

You may also have several mentor texts available for students to refer to. However students decide to add content-specific words to their pieces is up to them. Having a repertoire of suggestions to support your students' writing will make today's conferences a success!

After your students have decided on the words or definitions to revise, some children will need support deciding how to insert the scientific language. Having paper strips, Post-it notes, tape, or additional pages nearby always comes in handy and supports writers with revision. Although these tools are not new to your students, they may need some gentle nudging to use them.

Remember that in addition to supporting the work of the minilesson today, you'll also want to support children in general in finishing up their lab reports from the previous session, and potentially revising all their lab reports based on any of the learning they've had in the unit so far. They might be revising results pages based on what they've learned goes there, they might be adding more information about friction to their early lab reports, and so on. That is, your conferring today might steer children to recall that the work of the day is not based solely on the minilesson of the day—the work of the day comes from what a writer needs to do next and draws on all he or she knows!

MID-WORKSHOP TEACHING

Describing Scientific Processes with Familiar Words

"Writers, eyes and ears on me for a minute. Jesse just found a new word that we can add to our 'forces and motion' word bank: *speed*. He wasn't sure at first if that word belonged in the bank because it's a word that people use in everyday life, too, not just in catapult experiments. For example, if you're biking downhill, you go with a lot of speed. But then he realized that some of the other words in our bank are also everyday words, like *push*, *pull*, and *balance*.

"So I just want to be sure you all understand that while it's true that scientists use words that people outside their field rarely use (and sometimes don't even know), like *gravity*, they also use words that the rest of us use all the time. The difference is that they use these words to describe scientific processes. So when you come across a word like *speed* that you find yourself using over and over to talk about how an object moves, you're probably onto a word that can be added to our bank. Come on up here, Jesse, and add *speed* to our bank. And everyone else, if you find words that you think might fall into this category, let me know. We want to grow our forces and motion lingo so that our reports can become even more scientific and precise."

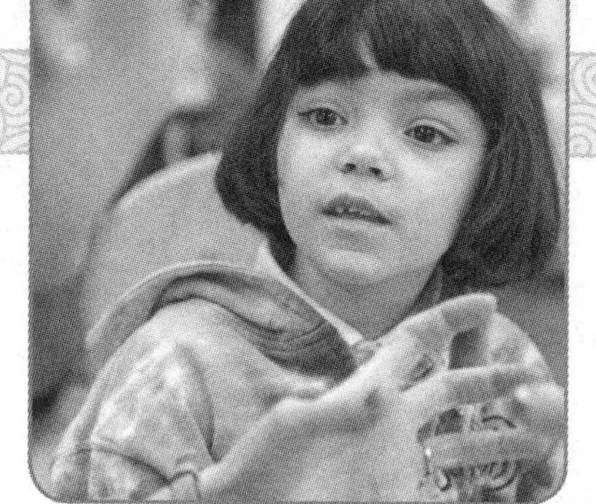

SHARE

Generating Information Book Topics

Celebrate children's growth as informational writers of lab reports through a museum share.

"Writers, in a short time, you have become serious science writers! You have learned not only how to record information but also how to grow and explain ideas as scientists do. When people visit an art museum, they move from painting to painting, studying each work of art to notice what makes it special. Today, you will do just that, but instead of looking at art, you'll look at one another's lab reports. You and your writing partner will read and study a few lab reports, noticing what's special in each one. You can even write compliments to the writers on Post-its!"

Ask children to brainstorm topics in their lives that relate to forces and motion to help them prepare to choose a topic for their information books.

"Writers, earlier I told you that scientists use words that everyday people use, too. Lots of the things that scientists study in a lab are things that people experience in their regular lives all the time. Jesse realized that if he went downhill on his bike, he would go down with a lot of *speed.* And when I asked all of you to talk about PE class using our forces and motion words, you had an easy time inserting words like *push*, *pull*, and *balance* to talk about a ball game. When you stop to think about it, each one of us is involved in real-life forces and motion activities all the time!

"Starting tomorrow, all of you are going to pick forces and motion topics that you'll be writing about in information books. You're going to think about how things work—the way things move—and use your knowledge of scientific process and writing to come up with questions about the world of movement.

"So right now, let's think together about examples of motion and forces we encounter in everyday life. Let's start by taking a mental walk through school. Think about the places you go to in this building and what you do here during a day. Thumbs up when you think of something that involves some motion . . . or force."

Henry's thumb shot up. "The doors!" he called out gleefully. "You have to push and pull them to make them move."

Nina said, "The slides and the swings have lots of motion. I mean, when you slide down, that's like gravity pulling you down, and when you push the swing, you can push it harder to make it move faster or higher."

Stella added, "When we dance in music class we have to push our bodies and spin around. It's like we make our own force."

"Yeah, and so does skiing," Henry said. "And the Slip 'N Slide, it's slippery and smooth and you move really fast."

"Wow, writers. All of the things you mentioned use forces and motion. As you go through your day, keep thinking about other everyday activities in this category. Sports are a great start, but see if you can push yourselves to think beyond that. Ask yourselves, 'What other kinds of things that I know about or do in life rely on force?' Off you go!"

Writing about Forces and Motion in Information Books

BEND III

Session 12

Drawing on All We Know to Rehearse and Plan Information Books

IN THIS SESSION, you'll teach students that writers choose topics they know a lot about and are experts on to write information books. Before writers write their information books, they plan how their information will go.

GETTING READY

- Writing center stocked with staplers and stacks of single sheets of paper for students to compile information booklets
- A baton (or pencil) to conduct the students in sharing their topic ideas (see Connection)
- Your own forces and motion–related topic and corresponding subtopics to share with the students (see Teaching)
- A couple of mentor tables of contents for students to study (see Share)
- Copies of blank tables of contents pages in the writing center (see Share)

Common Core State Standards: W.2.2, W.2.5, W.3.4, W.3.5, RI.2.1, RI.2.5, SL.2.1, SL.2.4, SL.2.5, L.2.1, L.2.2, L.2.3

TODAY MARKS AN IMPORTANT TURNING POINT in the unit. Although you have probably called your students "forces and motions experts," the fact is that up until now they have been functioning very much as learners. They've learned new ways to think, new structures in which to write, a new lingo for communicating—and they've worked largely on the same or similar experiments. Today, all that changes. Today students will assume the role of expert—and they will each go their own way. Today, you'll invite children to apply all that they've learned over the course of this unit to topics of their own choosing, topics about which they know something and can teach others.

Whereas writing information books, or "teaching books" as they were called in prior years, won't be new to students, the ways in which they structure these—and the new scientific approach they now bring—will be markedly different. By now students have a felt sense for how scientists pose and test questions or hypotheses, how they take measurements and record results, and how they make sense of their results, sometimes considering new possibilities. During this bend, students will draw on everything they learned in kindergarten and first grade about how to write a book that tells all about something, and they will draw on what they have learned during this unit about forces and motion to teach even more about their topic. As children construct books about ice skating, skateboarding, dancing, and so on, you'll steer them to incorporate all that they know about scientific writing, as well as about forces and motion.

In this bend, children will focus on the various sections of an information book as well as an assortment of features to best categorize and teach their topics. They will pay special attention to crafting introductions and conclusions, to making comparisons, and to assessing themselves as they write.

On this first day of the final bend, you will remind children of ways to rehearse for an information book, and they will select topics and get started right away, rehearsing by organizing across their fingers, sketching, and teaching others.

MINILESSON

Drawing on All We Know to Rehearse and Plan Information Books

CONNECTION

Drumroll the start of a new bend, and channel writers to quickly locate a topic they can teach an information book about forces and motion with.

"Writers, this is a really exciting day because we start a whole new kind of writing related to science—we'll be writing information books! And, just like in first grade, you will need to choose a topic you are an expert on. Since we are writing books about science, about forces and motion, you'll need to choose a topic that is about movement. It might be pinball playing, martial arts work, skateboarding, biking, riding scooters, dancing, swimming, or doing gymnastics.

"Most of you started thinking about your topics yesterday during our share session, so you might already know what you want to write about. But for those of you still deciding, ask yourself, 'What do I know that I can teach others? Is there a sport or a game or anything involving movement that I know a lot about?' Take a moment to think. When you think of something you know a lot about that you could teach others, thumbs up." Rather quickly, thumbs popped up across the room.

Ask some children to share their topics, in this way raising possibilities for children who still haven't selected one.

"Remember when I pointed my baton at you the other day and you named the experiment ideas you had? Well, today, when I tip my baton," I held up a pencil, "toward you," I arched it and pointed toward one child, "say your topic, loud and clear. Ready? Give a quiet thumbs up if you have your topic in mind." I sat tall, back arched, and said in a formal presentational way, "Information Books, by Room 103." Then I tipped the baton, and children sang out topics, one by one: scooters, roller skating, ice hockey, skateboarding. I ended by saying, "And many more," then brought my baton to rest, leaning in to the class.

Name the teaching point.

"Today I want to teach you that to write information books, writers might rehearse by talking, sketching, and then teaching people about their topic. Then, writers can use what they learn from sketching and teaching to help them revise their plan and write their texts."

◆ COACHING

Hopefully many of your students will have a good idea of what they'll write about before today's lesson begins, and you'll be able to move them rather quickly into the minilesson to give them as much writing time as possible.

TEACHING

Name and explain your topic choice.

"Writers, you all got me thinking about what I know a ton about. I'm going to try something a bit different and write a book about . . . *cleaning*! Yeah, I know you all are probably thinking, why did my teacher pick this topic? I don't tell this to everyone but, I like to clean. I think this is a good topic for me because it is something I do all the time, I can teach others about it, and when you clean there is definitely movement."

I partially chose this topic because I want to show students the wide range of topics available to them. I also didn't want to choose a topic that was on students' minds. I don't envision many seven-year-olds planning to write an information book on cleaning!

Demonstrate planning how your teaching (and writing) will go.

"I want to remind you that before you write an information book, you plan. And guess what? You *already know* how to plan! You plan by saying your information across your fingers, then quickly sketching or jotting notes about what you'll write on each page. Remember, writers, you will plan your information book almost the exact same way you planned your Small Moment books: think of an idea, say your information across your fingers, sketch (or jot notes), then write.

"Let me show you how I could plan my information book about cleaning." I lifted up one finger and said, "I could write about the tools I use to clean. There could be a part on using sponges or rags to wipe surfaces, and another part about how I use scrub brushes to remove things that are stuck. I could write about the friction in bristles—how you need friction to make cleaning work. That would be interesting to readers! Okay, let me take a page from my booklet and write *Cleaning Tools* as the heading." I wrote the heading and quickly sketched a sponge, scrub sponge, and scrub brush in the picture box.

"What could another section be?" I held up another finger and said, "Hmmm . . . . I could write about mopping the floors. I could teach people how to properly mop. You can't just slide the mop. You need to push it against the floor. Let me make that heading and sketch the steps of mopping the floor." I quickly demonstrated this.

"What else could I teach in my book?" I held up another finger and said, "I know! I could write a section about hand-washing clothes. I could write about getting stains out by rubbing them out or using a little brush to scrub them." Again, I wrote *Hand-Washing Clothes* as a heading and sketched how a person would do the job. "That could be at least three sections of my book! This looks like a good book for me to write. I have so much in my sketches, and I know I will be able to add this into my writing."

Very often, young writers begin an ambitious project only to discover that they lack the information necessary to write such a book with great detail and elaboration. When students orally plan and sketch their writing, they discover rather quickly whether or not the topic they've selected is one worth pursuing.

Name what you have done in a way that is transferable to another day and another topic.

"Did you see what I did? Did you see how I thought of a topic that included movement and that I was an expert on? Then I planned possible sections across my fingers and quickly sketched each section. I sketched or added notes to each page of my information book to get ready to write!" The students nodded their heads in approval. "And did you notice how I listed out loud the things that would go in each section before I talked about the next part?" Again, kids nodded.

ACTIVE ENGAGEMENT

Channel children to think of a topic they could teach others, then ask partners to have a go at describing each section of their booklet to each other.

"Now it's your turn to give this a try. You are going to take turns saying your sections across your fingers, then telling your partner how each section of your booklet might go. Partner 1, you are going to begin. Partner 2, you are the student, so adjust yourself so you are ready to learn. Partner 1, when you are in position, you can start." The room erupted into a chorus of voices, and I moved from partnership to partnership listening in. I coached into children's conversations, voicing over compliments.

"Henry, I love how you are touching a finger each time you teach something new. Alexsa, that is amazing—you are teaching your partner how the springboard for the vault in gymnastics is like the catapult! Keep going.

"Rebecca, you were saying your information across your fingers and adding details when something wasn't clear—wow! And Gabriella, you considered what your partner knew about your topic and used a teaching/explaining voice as you taught her about soccer. That means that when you go back to jot notes and then write, you will be thinking about your readers the whole, entire time. Well done!"

LINK

Restate the teaching point, making it applicable to not only today but every day.

"Professors, let's come back together. Wow! I am so amazed at all of the things that you know—it's like I've become smarter just by listening in to what you had to say. I learned about skateboards and gymnastics and soccer and flying kites. You all know so much! Now it's your chance to write an information book on the same topic that you have just taught to your partner. You can use your notes and sketches to help you!

"And after this, for the rest of your life, always remember that one way writers of information books get ready to write is they plan across their fingers, then sketch or jot notes on each section, and then teach someone. It can help to use a teaching voice and fingers or pages to organize information so that you teach one thing, then another, about a topic. Put your thumb up if you feel ready to write. When I see you are ready, I'll send you off so that you can get started sketching or jotting notes on your topic."

You may have felt that this unit was progressing unusually in that students have more often been working on the same process as is usually not the case during writing workshop. You'll really begin to see the benefits of the previous lessons' work, however, as students begin to branch out with their own topics, at their own pace—but with a solid foundation.

CONFERRING AND SMALL-GROUP WORK

Coaching Writers to Choose Content-Based Topics

AS YOUR STUDENTS BEGIN WORKING INDEPENDENTLY TODAY, you will notice that some children need support planning their angled information books. Students have a wealth of information regarding their topics. Your goal is to support them to plan and write these texts with focus, with information embedded, and with a sense of how their topics relate to movement and force.

As you confer today, you may want to carry around a list of vocabulary words from the bank of "forces and motion" words that you and students created together in Session 11. Students who are forgetting or struggling to write about their topics with attention to forces and motion will benefit from reminders of the related words. (You can use this strategy whenever students are embedding content into writing.)

If your students need a high level of support, you may choose to use demonstration as your teaching method. If your students do *not* need as much support, you may use guided practice as your teaching method. If you choose guided practice, during the teaching portion of the conference you may say, "Can I show you what I used to help me think about what information to include in my book?" You could pull out a list of vocabulary words from earlier in the unit, then say, "Let's reread this list together thinking about how these words fit with our topics." Point to each word as you read, pausing after each one to think and reflect. "Hmm . . . are you thinking about how words like *force*, *push*, and *pull* might fit with your topic?" Pause and look at the student approvingly, and give him some wait time. Be careful not to keep questioning the student, and try to keep your conference conversation-like.

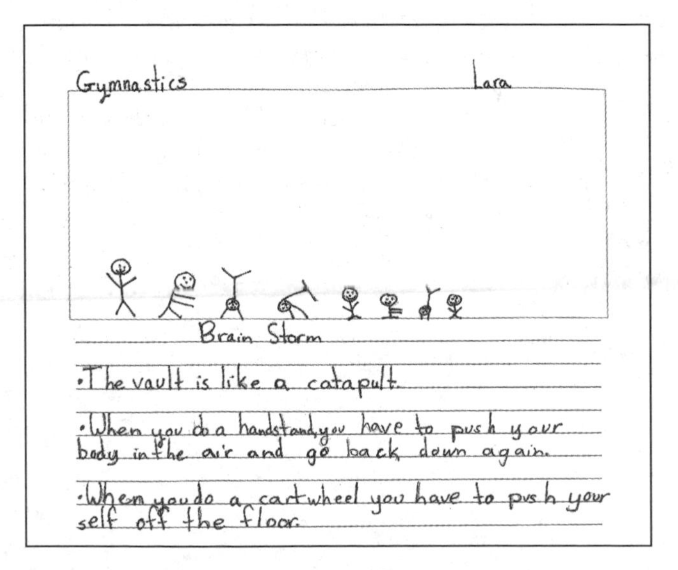

FIG. 12–1 Lara's brainstorm page for gymnastics

MID-WORKSHOP TEACHING

Using Teaching as a Rough Draft for Writing

"Professors of information books, can I interrupt you for just a minute?" I waited.

"Are you thinking about how teaching your topic can be a way to draft and revise how your books might go? The whole point of doing this teaching is to listen to yourself doing it and to think, 'No, that's not the best way to begin,' and 'That really works well. I should write it just that way!'

"So before you go on, will you all talk over the teaching you have heard? Start with the most recent teaching, and think of that teaching as a rough draft. The writer-teacher needs to ask, 'Did the topic seem like a good fit? Do you feel like this will work as an information book?' And listeners, will you talk over the teaching? What did the teacher say or do that was especially interesting, or that really helped you learn about the topic? What suggestions do you have? You can decide whether you finish some teaching first and then talk, or whether you start by talking. Go!"

SHARE

Crafting Tables of Contents

Recap that writers of information books make tables of contents that have a clear structure, then send writers off to do this.

"Writers, please join me in the meeting area with your work and pick up a blank table of contents page." As students assembled, I said, "Now that you have some plans for your books, it is important to revisit your headings and sketches to think about how you want your books to go. Making a table of contents will help take your plan to the next level!" I displayed a sample table of contents.

If you do not have a document camera, you can use an overhead projector or choose a nonfiction big book. You will want students to notice that there is an introduction, a series of chapters, and a conclusion. Pick your texts accordingly.

"Look at how this table of contents is organized and think about its parts. Turn and talk to your partner about what you notice."

After a few minutes, I continued, "Now I am going to put up another table of contents. Think about how it is organized and also about how both tables of contents are the same." I displayed the second table of contents and gave students a few seconds to read. "Thumbs up when you have an idea. Turn and talk to your partner."

When I had elicited students' attention again, I said, "Writers as I listened in just now, I heard you list these similarities between the two tables of contents. You said that each one begins with an introduction, that this is followed by a few chapters with different kinds of information, and that each one ends with a conclusion. Those are important structural similarities, and they are going to help you as you write your own tables of contents. They will be like mentors.

"Look back at your writing from today and think about all of your chapters and how you may organize them just like our mentors. Thumbs up when you feel like you have a good plan." I scanned the room and waited until students looked ready. "Now turn and share your plan for your book's table of contents with your partner."

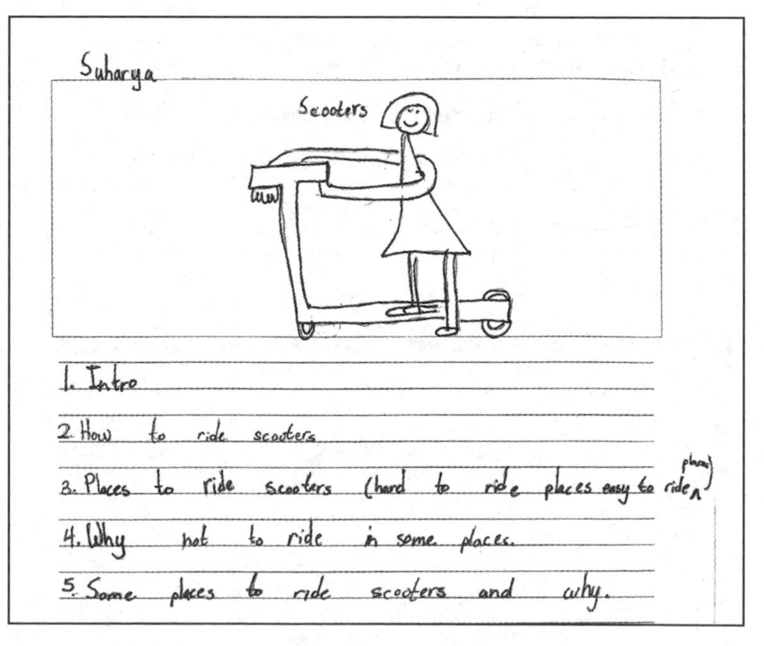

FIG. 12–2 Suharya's table of contents for her scooters book

Session 13

Tapping Informational Know-How for Drafting

IN THIS SESSION, you'll teach students that writers draft the chapters of their books by looking back at their tables of contents and their plans and deciding what they will write first, then next.

GETTING READY

- ✔ Students' booklets from the previous session and a writing tool (see Connection and Share)
- ✔ Your own demonstration information book pages planned with headings and sketches for a couple of your chapters (see Teaching)
- ✔ A marker or pen to draft a chapter of your information book (see Teaching)
- ✔ A student's sketches of his or her information book plan (see Active Engagement)
- ✔ Writing center restocked with copies of the Informational Writing Checklist, Grades 2 and 3. Have as many as needed for each student to refer to in today's share.

MANY OF THE BOOKS young children enjoy reading are expository texts. These books aim to teach, explain, and describe specific, focused topics to readers. In first grade, when students first learned to write informational texts, they studied how published informational books differ from narrative stories. They then planned their own teaching books, using their fingers to talk through their own topics, explaining and describing as they went.

Now, a year later, you are revisiting some of these tried-and-true strategies for teaching kids to organize and draft their writing chapter by chapter. They will be writing information books on topics of expertise but specifically tapping into the content that they explored in the previous bends. That is, you will challenge your students to synthesize all that they know about informational writing and all that they've learned in the previous bends about forces and motion.

In the previous session, students planned their information books, drafting possible tables of contents, along with headings and sketches across their booklets. It is now time for them to flesh out that structure with important facts and details, chapter by chapter. First, your students can revisit their tables of contents, headings, and sketches. Then, they can use those jottings to help them envision and then write the details and information that will fill the pages of each chapter. You may decide to guide your students to rehearse out loud independently or with a partner. Explain that information writers revise on the run, as they rehearse and draft their chapters. Revision, at this point in second grade, should not be an afterthought!

COMMON CORE STATE STANDARDS: W.2.2, W.2.5, RI.2.5, SL.2.1, SL.2.2, SL.2.4, SL.2.5, L.2.1, L.2.2, L.2.3, L.2.6

MINILESSON

Tapping Informational Know-How for Drafting

CONNECTION

Ask students to review their tables of contents, selecting a chapter they are especially ready to write.

"Writers, bring your booklets and a pen or pencil, and hurry over to our meeting area. When you're settled, reread your tables of contents and star the chapter you are just bursting to write!"

After waiting a moment for students to do this, I continued. "I'm glad to see you're all ready to write. I was so excited about what I'm planning to write today that this morning I got up extra early—I had an avalanche of ideas even before my alarm went off—and I jotted down my ideas for each chapter in my information book. I can't wait to share what I've written with you!"

◆ COACHING

You will notice that this connection is brief and efficient. We want the bulk of today's session to be reserved for drafting, so we aim here to quickly set students up for a productive day of writing.

Tyler

Table of Contents

1	Introduction	P1
2	How You Play *	P2-P3
3	Control the puck *	P4
4	Protecteve Aquipment	P5
5	Glossary	P6

FIG. 13–1 Tyler's hockey table of contents

Rebecca

Table of Contents

1	Basic Ballroom	2
2	Cha Cha - Basic Step	
3	Cha Cha - New Yorker	3-4
4	Cha Cha - Hand & Hand & Spot turn	5-6
5	Samba Steps	7,8,9,10
6	Jumpy Jive Steps	11,12,13,14
7	Rumba Steps	15,16,17,18

①

FIG. 13–2 Rebecca's ballroom dancing table of contents

Gabby

Table of Contents

1	Introduction	1
2	Basket ball	3
3	soccer	4
4	differences	5
5	similarities	6
6	why you need to do these things.	7

FIG. 13–3 Gabriella's soccer vs. basketball table of contents

Name the teaching point

"Writers, today I want to teach you how to use your quick sketches and plans from yesterday to help you draft your chapters. One way you can do this is by rereading each heading and looking at each sketch, imagining the words you will write. Then, you will write, write, write!"

TEACHING

Demonstrate planning and writing chapters.

"Remember yesterday I decided that since I am an *expert* on cleaning I would write an information book about that topic that includes all we have been learning about forces and motion. Then, I planned the chapters of the book across my fingers and quickly sketched each chapter." Students nodded. "Some of you started to write a section of your own information books yesterday, and I thought that I could give you a tip or two that will help you as you continue drafting your chapters.

"Let me show you how using my sketches will help me plan how my chapter about cleaning tools will go. I'm going to look back at the section I started and look at the sketches I made. Okay, I sketched a sponge, scrub brush, and scrub sponges," I said as I slowly touched each sketch. "Hmm . . . maybe I can start by writing about cleaning tools in general, and then I can get into detail, naming out when to use each tool. I could also weave in some science and some technical terms, and talk about how the smooth tools are good for wiping and do not produce a lot of friction and how the rough tools produce a lot of friction. Yeah, that sounds good."

In front of the children, I began to write:

> There are a variety of different cleaning tools that are excellent to clean surfaces with. Each tool serves a slightly different purpose and can be used in different areas of your home. One important cleaning tool is a sponge. Sponges absorb liquids and are soft. They are great on delicate surfaces and don't produce a lot of friction. Another cleaning tool is a scrub brush. The scrub brush has hard bristles that produce a ton of friction and are great to rub out dirt.

Restate the strategy in clear and explicit language.

"Writers, I am going to stop writing here for a second. I think you get the idea. Did you see what I just did? Did you see how I revisited each sketch to write my chapter? Then, I thought about what I could say about all of the sketches. Finally, I described each part in detail, weaving in some of the science work that we did."

ACTIVE ENGAGEMENT

Set students up to plan a chapter of a second-grader's information book.

"This is exciting but challenging work, writers. I know a second-grader, Daniel, who just like you is planning to write an information book. Daniel's book is about dogs. He has already decided what he is going to write about." I listed the topics across my fingers as I spoke: "Types of dogs, parts of a dog, and how to take care of a dog. Daniel couldn't be here today, but he asked me if we could help him think through one of his chapters, 'How to Take Care of a Dog.'

"What do you say, writers? Can we help Daniel write this chapter?" The kids nodded enthusiastically. "Great! Let's look at his sketches." I pointed to a sketch of pouring the food into a bowl, one of using a leash to walk the dog, and another of brushing the dog's fur. "Think about the sketches and what you could say about all of them as a whole. Then, think about what you could say about each one of them individually. Remember, weave some science in if you can. Turn and talk to your partner." I let partners talk for a few minutes—but not too long—about this chapter as I listened in. Soon, the class had created this chapter for Daniel's book:

> Taking care of a dog is a lot of hard work. You need to feed, walk, and take care of the dog's body. Dogs need fresh water in their bowls at least twice a day. You also want to feed the dog a mix of dry and wet food at least once a day. You will need to take the dog out for a walk so he can do his business at least three times a day. That's a lot of walking. Make sure you find a good leash for walking your dog. Dogs will pull away when they are walked, and you may have to pull them back to you, so you need a strong leash. A good, strong brush is best to use on a dog. The friction will help to untangle its fur.

Debrief—highlight the work students did on the sample chapter that is transferable to other books and other topics.

"Writers, it's so great to hear you thinking about what Daniel's readers will want to know, what they *need* to understand! You were thinking about the big idea of the chapter. Then, you started to write details. This chapter teaches readers about taking care of a dog, but the same work you did on Daniel's book you can do on your own books. First you thought about the sketches and about the chapter as a whole, and then you gave details about each part, including some of the scientific information you have been learning. That's good work."

It can be particularly motivating to students to study the work of another second-grader, which is why we chose to use Daniel's work here. You may decide to use the work of a previous student or a student from another school.

You will provide a varying amount of coaching during this process, depending on the needs of your class. Some classes may need more support from you early on in this generation process. Some teachers ask students to turn and talk during this part of the session and then elicit draft content from individuals. Some teachers start by drafting the opening sentences themselves. You'll want to do what makes the most sense for you.

LINK

Send students off to begin drafting their information books, tucking in reminders about how to write informational texts and how to connect their writing to the science they have been learning.

"I know that yesterday you all planned aloud with your partners, telling across your fingers, sketching across pages, and even beginning to write. Now as you go back to your tables, you may want to begin by taking a look at the chapter you have starred. Really look at your sketches and think, How will this chapter go? What can I say about all of the sketches and what can I say about the chapter as a whole? Then think, What details can I include for each part? What scientific information can I include? Then, write, write, write!"

CONFERRING AND SMALL-GROUP WORK

Asking Questions to Support Writing More

TODAY, YOU MAY DECIDE TO PULL TOGETHER A SMALL GROUP of writers who need to increase their volume—ones who have less than four sentences on a page. As you meet with this group, you can demonstrate how you can reread a chapter of your own writing, asking basic questions—Who? What? When? Why? How?—and letting those questions prompt you to add new information to your writing. Remember, the demonstration should be quick so that you can leave most of the time for students to practice the strategy in their own writing! As students are practicing, you might realize they would benefit from having these questions in written form. If so, create a note card with the question words on it, as a set of prompts for students to use until they outgrow it. As you wrap up your small group, you may want to remind students that they can use this strategy whenever they are writing information pieces with great detail.

You may also decide to pull together a small group of writers who spend a great deal of time on their diagrams and pictures—at the expense of their writing! Some children may be developing ideas related to their topic, but then are putting down those ideas solely in graphic form, not yet in writing. Others may be outright illustrating, with no thought to the relation of their pictures to the content, perhaps intricately designing the dress of the ballroom dancer in the "Jumpy Jive Step" chapter of a ballroom dancing book (see Figure 13–4). In this case, as you are affirming that writers of information books do indeed include both pictures and words, you may find it helpful to coach writers to act out a bit of their chapter, saying aloud with exact words what they are doing so that they can immediately turn to writing down those words.

MID-WORKSHOP TEACHING **Encouraging Fluency**

"Writers, remember to choose a chapter that you'll be able to pack with information. I started with the cleaning tools chapter because I had tons of information about those. It may be difficult to start with an introduction or a conclusion. You will probably want to start with the real meat of the teaching book, teaching all you know about your topics. Two sentences just isn't enough! So, if you're stuck, skip to a chapter whose knowledge is just pouring out of you. It might help to look at your chapter topic and think, What action could I explain in this chapter?"

Jumpy Jive Steps - Beginning: Order of Dance

Begin   move   Lift Pose

In the jive there are lifts, twirls and turns. In the jive the first step is the rock step. In the rock step you have to step back, bend your knees and then lift your front leg. The middle step or 2nd step is triple step. As you do triple step you lean the top of your body to the side of you and then dip with your knees after that you do three small steps to the right or left. You can never do a rock step on the left triple step. The lift comes in the middle or end of the dance: jive. You may get to slide back to your partner after a lift.

(11)

Jive: Rock Step

In the rock step you must lift up your knee in front of your body and keep your back leg straight and down.

Back Foot

Your back foot must be turned out halfway. The back leg must be straight and your foot should face the curtain on the stage. Before you go back into the triple step you should straighten your back foot.

(12)

Rock Step (continued)

Front Foot

You must put your foot down before you do a under arm turn, lift, in & out or triple step. It will not be smooth when you do the next step.

(13)

Jumpy Jive Steps–Beginning: Order of Dance

In the jive there are lifts, twirls, and turns. In the jive the first step is the rock step. In the rock step you have to step back, bend your knees, and then lift your front leg. The middle step or second step is triple step. As you do triple step you lean the top of your body to the side of you and then dip with your knees, after that you do three small steps to the right or left. You can never do a rock step on the left triple step. The lift comes in the middle or end of the dance: jive. You may get to slide back to your partner after a lift.

In the next step you must lift your leg in front of your body and keep you back leg straight and down.

Your back foot must be turned out halfway. The back leg must be straight and your foot should face the curtain on the stage. Before you go back into the triple step you should straighten your back foot.

You must put your foot down before you do a underarm turn, lift, in and out, or triple step. It will not be smooth when you do the next step.

FIG. 13–4 Rebecca's Jumpy Jive chapter

SHARE

Self-Assessment and Goal Setting

Set one writer up to read his informational text aloud while the class evaluates that text to see ways he does (and does not yet) show he is meeting the criteria for good informational writing.

"As you gather for a share session," I said, "take one of your Information Writing Checklists for assessing your own writing." The Information Writing Checklist, Grades 2 and 3 can be found on the CD-ROM.

"As I walked around today, I noticed how many of you are already writing in ways that show you know a lot about writing like scientists! Tyler has finished his information book, and he wondered if I could help him check that he has done everything on the second-grade checklist for writing information books. I thought maybe we could *all* help Tyler. In a minute, Tyler is going to read his book aloud to us and display it on the overhead, so that we can help him find places where he's writing like a scientist!

"Let's all take jobs. Who wants to see if Tyler taught his readers some important points about his subject?" Some children volunteered. "And who will listen to see if he named his subject and tried to interest his readers? Who can listen for that?" Soon each item on the checklist was taken. After Tyler read his piece aloud, the class agreed he had done many of the items on the checklist, but that he still needed to group his information into different parts.

Information Writing Checklist

	Grade 2	NOT YET	STARTING TO	YES!	Grade 3	NOT YET	STARTING TO	YES!
	Structure				**Structure**			
Overall	I taught readers some important points about a subject.	☐	☐	☐	I taught readers information about a subject. I put in ideas, observations, and questions.	☐	☐	☐
Lead	I wrote a beginning in which I named a subject and tried to interest readers.	☐	☐	☐	I wrote a beginning in which I got readers ready to learn a lot of information about the subject.	☐	☐	☐
Transitions	I used words such as *and* and *also* to show I had more to say.	☐	☐	☐	I used words to show sequence such as *before, after, then,* and *later.* I also used words to show what didn't fit such as *however* and *but.*	☐	☐	☐
Ending	I wrote some sentences or a section at the end to wrap up my piece.	☐	☐	☐	I wrote an ending that drew conclusions, asked questions, or suggested ways readers might respond.	☐	☐	☐
Organization	My writing had different parts. Each part told different information about the topic.	☐	☐	☐	I grouped my information into parts. Each part was mostly about one thing that connected to my big topic.	☐	☐	☐
	Development				**Development**			
Elaboration	I used different kinds of information in my writing such as facts, definitions, details, steps, and tips.	☐	☐	☐	I wrote facts, definitions, details, and observations about my topic and explained some of them.	☐	☐	☐
Craft	I tried to include the words that showed I'm an expert on the topic.	☐	☐	☐	I chose expert words to teach readers a lot about the subject. I taught information in a way to interest readers. I may have used drawings, captions, or diagrams.	☐	☐	☐

Channel all members of the class to transfer this work to their own books, rereading their own writing to check it against the same criteria that the class used to assess the one child's writing.

"Now, writers, lay the writing you did today on the carpet in front of you alongside the checklist." Once they'd done this, I said, "As I read off each item, will you check your own writing? When you hear an item from the checklist, if you can find it in your writing, put a smiley face near it on your page, and check that one off on the checklist. If that item is not yet in your writing, star it, and remind yourself to try that next time you write and revise your work." I read the writing goals one by one, leaving intervals for children to scan over their work for evidence of each item.

End with a note of celebration and resolve, reminding children that they can always use the Information Writing Checklist as a tool for improving their writing.

"How many of you found many of the things on the checklist in your writing?" I looked around. "And did some of you learn about things you can work on tomorrow?" All hands went up. "Great work! Remember that writers always set goals for their work, check their writing to see what they've achieved, and then make new goals for themselves. Tomorrow, before you start writing, look at your checklists again to remind yourselves about your goals for the day."

Session 14

Studying Mentor Texts

Integrating Scientific Information

IN THIS SESSION, you'll teach students that writers look at mentor texts to find ideas for their own writing. When studying information books, writers look to see how the authors integrate scientific information into their writing in a way that connects to their topics.

GETTING READY

- ✔ Several information book mentor texts displayed—perhaps ones from your classroom library's baskets. Tab pages of examples of cross sections, cutaways, transparent pages, zoom-ins, pictures from different perspectives, and the corresponding texts that explain the pictures; we used Stephen Biesty's *Incredible Cross-Sections*
- ✔ Your own teacher-made information book—Make a copy without revisions, and then have two ready-made versions with the revisions added in; revisions should show how to add a "science voice" to a part of your book and how to show movement with arrows to demonstrate forces and motion (see Teaching and Active Engagement)
- ✔ "To Put More Information in Informational Writing . . . " chart (see Share)
- ✔ Writing center replenished with paper and revision tools such as pens, strips, flaps, Post-it notes, staples, and tape

Common Core State Standards: W.2.2, W.2.8, RI.2.6, RI.2.7, RI.3.7, SL.2.1, L.2.1, L.2.2, L.2.4.a, L.2.5, L.2.6

IN TODAY'S SESSION you will remind your writers to look to mentor authors for ideas as they work to include science in their information books. While the energy of the last two sessions has been in driving kids toward drafting a teaching text, here the goal is to engage children again with the idea that they should be writing like scientists. You will want several examples of information books that incorporate aspects of scientific writing—some of which students might have already read when writing their lab reports. Diagrams like cutaways, cross sections, or transparencies will help students envision new ways to incorporate information into their own scientific writing.

You might want to create some fanfare around students' attempts to make their information books look and sound like "real" science writing. The kinds of strategies your writers will be incorporating into their work by the end of this session are likely to feel new for them, but the process of studying a mentor text will be familiar. Coach children to ask themselves, *What do I notice the author has done that I could also try?*—and ensure that the lessons they draw from this careful reading support their independence as writers. In this way, even as your students are just beginning to understand a new genre of text, they will be standing on the shoulders of what they already know, reaching new levels of achievement in informational writing. In this session's mid-workshop teaching, you will guide students to use definitions to develop their ideas. All of this work supports elaboration. Your hope for today will be to see your students writing long down the page, adding revision strips and flaps as they try out new things in their writing.

As students go off to work, it is important that they use the strategies they are learning with purpose. Students may need extra support to make connections between the topics they have chosen and the science content they have learned. Here again, you will want to remind children to see their texts not only as writers but as scientists. Getting your writers to ask *how* and *why* will help them see where scientific information fits into their writing, and studying the work of other authors will give them a vision of how to present that information meaningfully.

MINILESSON

Studying Mentor Texts

Scientific Information Additions

CONNECTION

Remind children of the path of their learning so far in this unit, and let them know how it connects to today's teaching point.

"So here's where we are in writing workshop. For the first part of this unit on writing about science, you were writing lab reports as you worked on experiments. That kind of writing is meant to help other members of the community understand what you've learned and how you learned it.

"But people write about science in other ways, too, of course. Science writing is a part of many, many informational books—informational books just like the ones you are working on right now!"

Name the teaching point.

"Today I want to teach you that when writers are trying out a new kind of writing, they often look at published writing to find examples of how it can go. Then they try it out themselves. In particular, today we will look at ways that writers of information books include scientific information in their writing."

◆ COACHING

Every now and then, a connection simply reminds children of the work they've been doing and its purpose in the world.

TEACHING

Tell students there are many ways science writing fits within information books. Explain that they can figure out some of them by looking at published texts. Then they can try out those ways!

"Last night—since we are writing information books related to science, and since that is a new kind of writing for me—I was looking through a stack of information books to see how these go when authors put scientific information into their writing. As you know, writers can always look at published writing to help them learn how to write in new ways.

"I was amazed! In only a few moments of looking, I found six examples of ways writing about science fits into information books. I think if you looked at information books this same way, you could find even more ways authors write about science in their information books. And here is the exciting part: even though these books are written by adult professional writers, I am quite certain that you all can use some of these same ways in your own writing!"

Point out a technique writers use to include science in an informational text.

"Each time I found a way an author included science knowledge in her information book to make it stronger, I thought to myself, 'You all could write like that! You have scientific knowledge and you have information books, and you could write like that!' Let me describe one example.

"Many of the books I looked at incorporated different kinds of text. Let me show you what I mean." I opened *Incredible Cross-Sections* (Stephen Biesty, 1993) and displayed it, pointing out the several different kinds of writing—the different sizes and styles—on each page. "You see how this paragraph explains the whole thing along the side, in kind of big letters? And then over here there are smaller words, positioned closer together that tell detailed scientific facts and histories and explanations, to go with the big picture?

Teachers, there are lots of popular examples of informational texts with distinct through-lines in them. Nearly all the DK Readers follow this pattern, as do the Cross-Sections books. Many books for children about animals include both a story about the animal as the main text and also facts about the species written in a different kind of font and type size.

"That gave me the idea that you could use this technique to incorporate scientific information into your own writing—you could use a different font or a different color and add in another whole voice to your books, one that talks about the scientific parts of each page. So you'd have written the main part of your book in one size and color, and then you could go back and add scientific information in little letters, or in big letters, and maybe in another color like green or red!"

Show students an example of your own writing that incorporates this technique and channel them to think how to do likewise in their own books.

"But then I thought to myself, could we *really* write that way? Because sometimes it is easy to think you can do something—easy to say you can do it—but much, much harder when you actually try to do it. So, I figured I had better try it out myself. Here is my page *before* I tried this technique (see Figure 14–1). Can you imagine how I could revise it in this way? Think for a moment to yourself."

I paused before showing them my own revision (see Figure 14–2). "Let me show you how it came out for me."

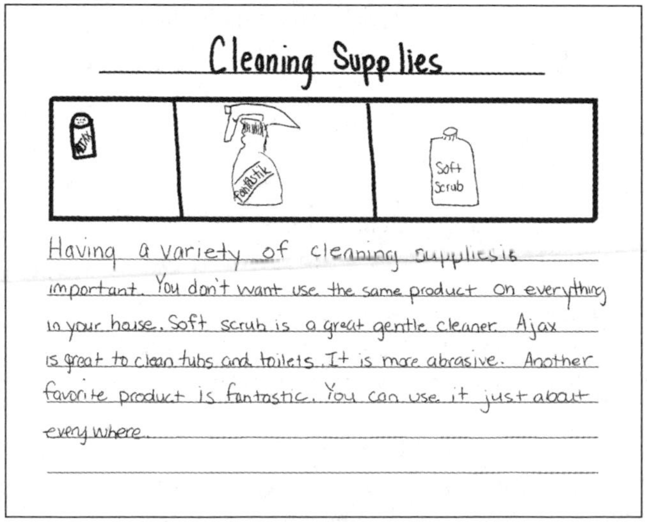

Cleaning Supplies

Soft Scrub

Having a variety of cleaning supplies is important. You don't want use the same product on everything in your house. Soft scrub is a great gentle cleaner. Ajax is great to clean tubs and toilets. It is more abrasive. Another favorite product is fantastic. You can use it just about everywhere.

FIG. 14–1 Teacher-made page from cleaning information book

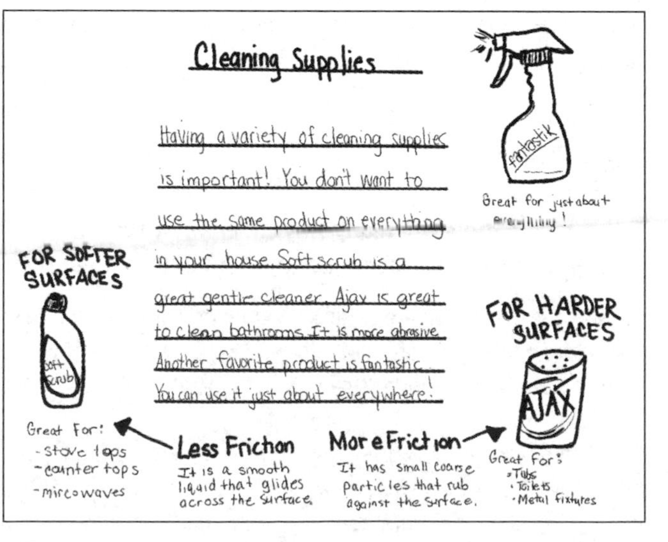

FIG. 14–2 Teacher-made page, now with some forces and motion, in a new color, style, or size

"So you can see, this technique helped me put scientific information in my book. I changed where I put my photos on this page, and I added the scientific words 'Less Friction' and 'More Friction' in big, bold letters and included how those words relate to my topic. Could you see this working for your own book? Turn to a page you've written or planned and tell your partner how you might use this technique on that page."

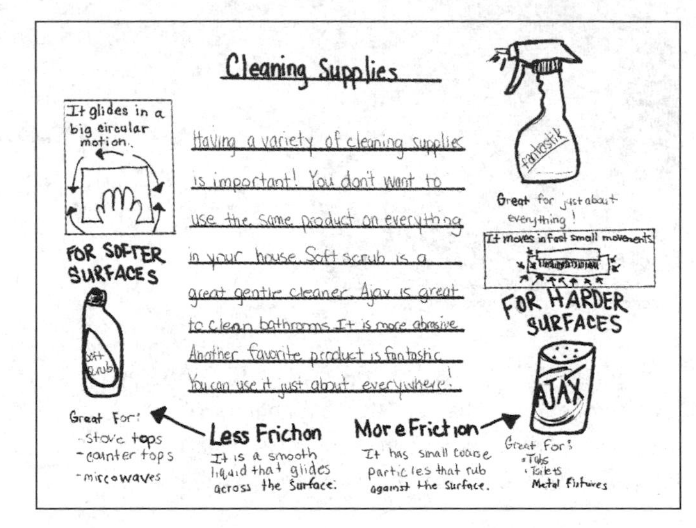

FIG. 14–3 Teacher-made page with arrows showing motion

ACTIVE ENGAGEMENT

Point out another technique, and ask students to help you figure out how to use it in your own writing.

"Did you get some good ideas? Now that you've seen how I studied a published book, got an idea, and put it to use, let's try it together. After this, I'll ask you to try it all on your own, with the information books that are in the centers of your tables.

"This is another technique I noticed: in lots of the books, the authors use arrows to show how something works! Labels aren't enough—the authors need to show movement and how one thing leads to another, so they use arrows to show forces and motion. Turn to your partner and think together to see if you can help me use this technique to say more on this page. Here it is again." I displayed my second page again and listened in as students talked.

After just a few minutes, I called students' attention back. "I heard so many great ideas for how to add arrows to my page to show motion! Many of you suggested I should add arrows to show the amount of friction. Wow—great idea! Here is how my revision for adding this technique turned out for me" (see Figure 14–3).

LINK

Remind writers that they know how to use authors as mentors, and ask them to get started finding, in published works, a technique that can help them with their current writing.

"Writers, remember at the beginning of the year we studied *Owl Moon* and other books, and noticed what the authors of those books did and why, and we imitated techniques when they helped us? This is that same kind of work! Now, you are looking at information books with the intention of finding a way to tell your readers about the scientific information that goes with your topics. Look through some of these information books to see if you can discover a technique that would be helpful to you. After you've found one or more, go ahead and incorporate it into your writing. And jot down the techniques you try so that we can collect a list of them later. Get started!"

Coaching Writers to Discover Connections to Science Content

AS YOU CONFER WITH YOUR STUDENTS TODAY, expect to see them progressing through their chapters, thinking about how best to organize each section. You may notice that some students could use support making connections between their topics and the science content they have been learning.

Many of the topics that students write about require a description of tools or equipment. One way to help students pop out science content in their information books is by nudging them to articulate *how* and *why* to use a particular tool or material. Imagine you are conferring with a student who is writing about gymnastics. In her chapter on the uneven bars she describes chalking her hands before her routine. You could support this writer by asking, "What are you planning on teaching your readers in this section?" and then encouraging her to think with you about the science behind her description of the bars and the chalk. You might ask, "Why do you put chalk on your hands?" Then when she says, "It helps my hands spin around the bar. My hands won't stick to the bar when they are chalky," you might teach her explicitly: "Wow! Look at that—the chalk reduces the friction, doesn't it?!"

Remember, however, not to steamroll your writers' agenda. Coaching students to ask and answer questions about their topics can help them to make connections to science content on their own and to find similarities between what they have observed in their science labs and what they observe in real life. Prompting students to consider the tools and equipment associated with their topics is just one way to encourage them to explore the world as scientists, as thinkers who ask *how* and *why*. As always, whether you are teaching minilessons, small groups or individuals, students should come away from conferences with transferable strategies that they can use whenever they are writing, not just when they are writing a particular piece. Your goal is always to teach the writer, not the writing.

MID-WORKSHOP TEACHING **Adding in Definitions and Tantalizing Readers with Science**

"Writers, can I stop you for a moment? Ana was looking at an information book *How Things Are Made* (Sharon Rose, 2003) that had a photograph of a lightbulb on a factory conveyor belt. The photo had this caption, 'Flames melt the necks of glass light bulbs. The filament—the part of the bulb that lights up—is sealed inside during this process. Turn the page to see how light bulbs are made' (31).

"This little caption gave her an idea of one more way she could incorporate scientific information into her information book: she is going to use those lines—dashes—to put in some definitions that will help readers understand the vocabulary of her topic, soccer. Here is what she wrote: 'When you are near the goal, you should kick it in, even if someone is partly blocking you. Don't turn around because you might get a deflection—when a ball bounces off someone—and so you might need to kick it in again!'

"The reading she did also gave her the idea to make a picture on her first page. Then, she'll make a caption that sets the reader up to want to read and learn more. So she drew a picture of a player throwing a soccer ball very far and wrote, 'To learn more about how to make a throw-in go far, turn the page!'

"By now you have definitely discovered from these published information books some ideas for how to put scientific knowledge into your own information books. I'm sure if you haven't already, you are ready to try some of these ways now!"

SHARE

Collecting Ideas from Mentor Texts

Collect children's suggestions of ways to integrate scientific knowledge into their information writing. If needed, add a few techniques of your own.

"Writers, you all have been finding and using all sorts of ways to make your information books even more rich with information, especially scientific information. I started a chart of what we've talked about so far. What else can we add?" Soon the class had come up with the following list.

To Put More Information in Informational Writing . . .

- Add a new voice in a different size or color.
- Use arrows to show how something works.
- Use dashes to add definitions.
- Add captions to pictures.

① Introduction

This book will teach you alot of things you need to know to play hockey. How many things do you put on? This book will answer that. Take this with you when you go ice skating because this book will teach you how to skate.

② How You Play

skates

repet

Skating-The first thing you have to do is check your skates on ice because they can be loose. Then you bend your knees. After that you use alot of force to push with one skate and put it back together. And repete, and repete, and repete on skate at a time.
Stoping- Keeping your knees bent, turn one skate sideways infront of you. It takes

③ alot of friction to stop you so you would turn your skate alittle before where you want to stop.

④ Control The Puck

no puck one hand

puck 2 hands

When you stop with a puck on your stick, you hold the stick with 2 hands because one hand is not enough strength. But with no puck you can stop with one hand. The time you don't need the puck is when your playing defence. You have to try to stop the other person so you slap his stick going backwards.

⑤ Protective Aquipment

For aquipment you wear a neck gaurd, gloves, helment, chest protecters, elbow pads, big long socks that go around your knee pads, and special pants. That will totaly keep you safe. My favorit aquipment is big, you wear it around your body and it goes under your shirt. What is it?

FIG. 14–4 Tyler's hockey book with science tucked inside

Session 15

Using Comparisons to Teach Readers

IN THIS SESSION, you'll teach students that writers use comparisons in their information books. They compare something that is new for their readers to something their readers already know.

GETTING READY

- ✔ Your own story or analogy that illustrates how using a comparison can help readers understand something unfamiliar (see Teaching)
- ✔ Teacher demonstration text with a few examples of comparisons embedded in a chapter (see Teaching)
- ✔ "To Put More Information in Informational Writing . . ." chart from Session 14 (see Link and Share)
- ✔ Writing center filled with fresh materials: single sheets of the new paper choice, extra table of contents papers, half sheets of paper, replacement pens, tape, and Post-it notes

Common Core State Standards: W.2.2, W.2.5, W.3.2.b,c; W.3.3.b, RI.2.6, SL.2.1, SL.2.3, SL.2.4, L.2.1, L.2.2, L.2.3, L.2.6

DONALD GRAVES HAS SUGGESTED that communicating clearly with an audience is a skill that takes a great deal of practice. In *A Fresh Look at Writing* (1994), he observes, "Young children often believe they will be available to supply the extra information their readers may need." (Understandably so—what grown-up author hasn't wished she could be present to defend her writing?) So in a sense, while this session gives students a new strategy for elaborating their ideas—writing comparisons—the session also affords your writers the opportunity to develop a sense of audience. You will ask them to anticipate readers' confusion or questions, and to try to clarify those things in writing.

You might encourage students to imagine that readers are learning the content in their writing for the very first time. Since there is so much new information, making comparisons to familiar things can be a powerful teaching tool. Students will be familiar with this strategy, but the aim in this context is new. Students are comparing not only to describe but also to teach—to provide readers with those aha moments when a new idea suddenly makes sense.

In the teaching portion of this session, you will illustrate using comparisons as a strategy for incorporating details that teach. Introduce this concept by conveying a straightforward example from your personal experience about learning something via a comparison. Then throughout the lesson, you will scaffold their understanding by giving them opportunities to be both readers (listeners) and writers of shared texts. As Graves suggests, shifting back and forth between the role of writer and reader strengthens a writer's ability to see what is missing or needed in her text.

As you move through this session, you will probably find that some students struggle to write apt comparisons—and that's okay. Making comparisons is a sophisticated craft move, especially when the aim is not just to describe, but to explain, too. The important work here is to nurture their understanding that writing—*their* writing—transcends them, and that writers make every effort to let that writing speak for itself.

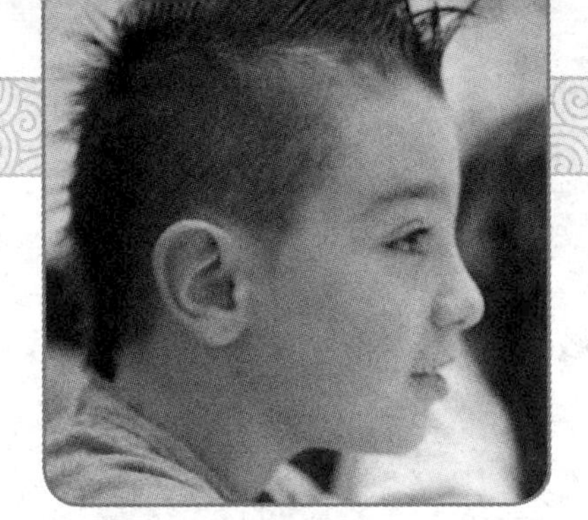

MINILESSON

Using Comparisons to Teach Readers

CONNECTION

Remind writers that they already know what it means to write with details.

"Writers, your books are really coming along, and you are writing in a way that teaches your readers so much about your topic. Yesterday, I was talking with Suharya about her book on scooters, and she reminded me of something very important for your writing.

"Suharya got to the chapter in her book called 'How to Stop,' and she was stuck. 'I don't know what to write,' she said. 'When you want to stop you just put the brakes on.'

"Like this? I asked her." I pantomimed scooting across the meeting area, tapping the brakes behind me with one foot and hopping forward not being able to stop. "Whoa!"

"Nooo!" the class called out.

Tyler offered, "That's so obvious! She should have said you have to keep your foot on the brake until you slow down, then stop."

"Good idea, Tyler. That's what Suharya told me, too. She knew all along what she wanted to write. She just had to remember to write with details. That's your tip, writers, always and forever. Write with details when you want to teach your reader. When you want to explain *how* to do something, tell how *much*, how *far*, how *fast*. You might think of your reader as learning about your topic for the very first time. There are lots of ways to include details in your writing, and today I want to focus on one of them."

Name the teaching point.

"Today, I want to teach you that nonfiction writers often use comparisons in their teaching books to show readers how the new thing they are explaining is similar to something readers already know."

◆ COACHING

Teachers, your students will have practiced writing comparisons before, but this will be a new context—and a new aim as well. Not only do comparisons here aim to describe, they also aim to explain, to teach.

TEACHING

Tell a story that illustrates how using a comparison can help readers understand something that is unfamiliar to them.

"Writers, can I tell you a story? The other day I was making a new dessert for the first time—taffy! The taffy recipe said 'Boil the mixture until it reaches the hard ball phase.' I had no idea what that meant! So I called my nephew—he loves to cook—and I asked him. He said, 'Sure, the hard ball phase, that's easy. When you put a drop of the boiled syrup in cold water it will get hard and round, like a marble.'

Teachers, by all means replace our story with an example from your own life if you prefer. We like this comparison because it typifies the type of explanatory comparisons we hope students will include in their own writing. Feel free to borrow it if you'd like!

"Writers, when my nephew said the syrup would look like a marble, I could picture just what he was saying! And nonfiction writers do that all the time. They compare something that is hard to picture (like boiled sugar syrup) with something that is easy to picture (like a marble)."

Show an example from your demonstration text of using a comparison to help readers picture a detail.

"Writers, let me show you how I used a comparison to help readers understand what I wanted to teach in my book about cleaning. I started off writing a detail, then I added a comparison to help readers picture it."

> Soak the mop in the bucket until it's drenched-heavy and dripping like long wet hair.

"Even if you have never mopped a floor before, this gives you some idea of how to start with the mop. Everyone can picture what long wet hair is like!"

ACTIVE ENGAGEMENT

Set writers up to try adding a comparison to a page from your demonstration text.

"Writers, could you help me add another comparison to my book about cleaning? After a weekend with lots of guests at my house, I have lots to clean. I want to explain to readers what a mess there is to mop up! My kitchen floor is covered with muddy footprints and crumbs. It's so dirty! Could you help me come up with a good comparison to describe this dirty floor to my readers? Turn and talk."

I listened as students talked about my dirty kitchen—their imaginings ranging from the fantastically hyperbolic to the literal. When I heard Sophia say, "Her kitchen floor is so dirty it's like dirt," I coached in.

"That's a great idea, Sophia! I love how you are using the word *like* to make a comparison. Let me help you make that comparison even stronger. Remember, making comparisons is a way of using lots of details. My floor is dirty, that's true! It has dirt on it. To make a strong comparison let's say that my floor is *like*, or is very *similar to*, some other thing that's dirty. Can you describe some other place that has mud and dirt and grime on it?" I waited as Sophia sighed thoughtfully. Finally she shouted, "Her floor is dirty like a muddy soccer field after it rains!"

You can choose any topic for your demonstration piece. As you are writing, be on the lookout for multiple places where the use of comparisons can lift the level of the piece. You'll need one part to use for your demonstration, and another for students to try!

"That's it—now you are using details to help readers picture exactly what you mean!"

LINK

Remind writers of all of the strategies they know to teach readers.

"Writers, I want to remind you that you know a lot of strategies now for adding more information to your books. Remember our chart we started yesterday, 'To Put More Information in Informational Writing . . .'? I'm going to add 'Make a comparison' to this chart. Anytime you are stuck or aren't sure what to add to your writing, you can look at this chart for strategies to teach your readers."

To Put More Information in Informational Writing . . .

- Add a new voice in a different size or color.
- Use arrows to show how something works.
- Use dashes to add definitions.
- Add captions to pictures.
- **Make a comparison.**

"As you go off to write today, remember that your job as writers of information books is to teach readers as much as you can about your topic. To teach well, you need to use lots of details. Today I offer you a challenge: see if you can come up with even more ways to make your chapters chock-full of great information than ever before."

Conferring to Ensure Students Have Grasped the Essentials of the Unit

AS YOU APPROACH TODAY'S CONFERENCES and small-group work, you'll want to keep the big picture in mind. Your last two sessions have spotlighted helping writers communicate clearly by writing with detail and incorporating craft. Just because today's minilesson taught that writers of informational texts often use comparisons to help readers clearly picture things, your focus need not be restricted to this during conferences. As you research your students' writing today, you may see them purposefully incorporating items from the "To Put More Information in Informational Writing . . ." chart. At the same time, you may discover that your students need support in other areas, so you will have to differentiate to meet their needs.

If you notice that some students still need more support elaborating, you may decide to coach them to review the repertoire of strategies introduced in the "To Put More Information in Informational Writing . . ." chart. As a writer reviews the chart with his piece of writing in hand, you may ask, "Can you choose a strategy that will help you teach more in your writing? It could be a strategy that you want to try and are a bit unsure of—I am here to help you!"

You also may notice that some students "try out" every strategy that you have ever introduced—you say, "Jump," and they say, "How high?" These writers need to learn to be more purposeful in their use of craft aiming to use it in ways that actually lift the level of their work. You could teach students to revise their writing by rereading and asking themselves, "Does this part fit here? Does it help teach more and make my writing stronger?" You might then coach writers to take out parts that do not belong. Every writer needs to reread to eliminate material that doesn't fit.

For children who do take on the work of the day's minilesson, you might suggest they work together to mine yet more lessons from the mentor text. Leave four kids with a mentor text and a pack of Post-it notes, and charge them with finding additional things the mentor text does that they could also try.

MID-WORKSHOP TEACHING

Using Your Senses to Include Details

"Writers, you are writing your information books to teach as much as you can about your topics. I love how you are writing and revising on the run and using the chart to purposefully expand your writing. I wanted to give you one more tip today about how to teach or describe as much as you can in your books. You can use your senses. You can ask yourself, 'What do I see?' 'What do I hear?' 'What do I smell?' 'What do I taste?' 'What do I feel?' Then, you can use this to add detail to a section of your book where you think it will help teach your reader."

SHARE

One Student's Work at Adding Details

Highlight one student's writing, showing how the addition of comparisons, examples, or other details helps readers understand what the writer is trying to convey.

After gathering the students in the meeting area, I said, "Writers, I am so impressed with how thoughtful you have been when writing your information books. You write a part, reread, and then revise on the run. You are thinking about all the craft and elaboration strategies we have studied and are thinking about making your piece the best it can be.

"Now I want to share with you a chapter from Farai's information book (see Figure 15–1). Farai used comparisons in his book on drumming. In this section, he is writing about changing the sounds of the drum. Listen for how Farai used a comparison and then how he also did something else to add more detail to his piece. I'm going to read it to you, and your job is to figure out what Farai did."

①
Big Sonuds and little Sonuds
little sound | Broom | Chik chik | Spring | Base | high Hat
There are many cins of drums. They have difint sisise and difnt souns. The bigets drum is the Base drum. That is the one That sits on the flor. You have to ues the petl to strike it. If step on the petl soffly it will

②
make a small sound that is still strog, it will sound like your hiting a puching bag. But if you step on the petl with a lot of force it can sound like THUNDER!

③
Farai
There is also the hi hat. These are two sibols that gosh togeter. They cind of sound like two metol pot cavers wehn you smash them togeter. You can hit the hihat with a drumsick or you can pedol them too. If you ues the drumstiks it souns like shingshing! But if you ues the pedls they sound like chik chik chik.

FIG. 15–1 Farai's detailed chapter on drumming

Chapter 1. Big Sounds and Little Sounds

There are many kinds of drums. They have different sizes and different sounds. The biggest drum is the bass drum. That is the one that sits on the floor. You have to use the pedal to strike it. If you step on the pedal softly, it will make a small sound that's still strong. It will sound like you're hitting a punching bag. But if you step on the pedal with a lot of force, it can sound loud like THUNDER!

There is also the hi hat. These are two cymbals that crash together. They kind of sound like two metal pot covers when you smash them together. You can hit the hi hat with a drumstick or you can pedal them too. If you use the drumsticks, it sounds like shring! shring! But if you use the pedals, they sound like chik, chik, chik.

"Now, writers, think about how Farai used a comparison, and also think about what else he did to add more detail to his piece. Thumbs up when you are ready to share an idea with your partner." I listened in for a moment and then called the class back together.

"I heard so many of you notice how Farai compared the sounds of the different drums—to thunder and to metal pot covers. I can just hear the sounds, can't you? I also heard some of you say that Farai used sound words in his writing—the shring! shring! and chik, chik. Great observations, writers! Farai used the strategy we talked about during the mid-workshop teaching to add detail to his piece and it worked so well to teach us about drums. Some of you may want to try what Farai has done in your own writing.

"I am going to add 'use your senses to make a description' to our 'To Put More Information in Informational Writing . . .' chart. Remember, you can use this chart to help you whenever you are trying to teach as much as you can in your information books!"

To Put More Information in Informational Writing . . .

- Add a new voice in a different size or color.
- Use arrows to show how something works.
- Use dashes to add definitions.
- Add captions to pictures.
- Make a comparison.
- **Use your senses to make a description (what do I see, hear, smell, taste, or feel?).**

Session 16

Showing Hidden Worlds with Science Writing

"THE MOST BEAUTIFUL THING WE CAN EXPERIENCE is the mysterious. It is the source of all true art and all science," said Albert Einstein. This lesson addresses the mysterious in concrete ways. In the beginning of this lesson, we introduce students to a modern-day scientist who noticed something no one had ever noticed before, made a hypothesis, tested it, and then went on to make new hypotheses and new discoveries. By talking about popular scientists, we help students connect their own experiences in science with adult scientists' work.

As you work through this session, you may want to highlight for students that the writing knowledge they gain today will live beyond this unit, beyond writing workshop, and beyond their year in second grade. They can always learn from experts in any field to make their writing—in this case, their informational writing—even stronger. Today you will tell students that studying and writing about science often means explaining things that are out of our range of perception, things that normally would be hard for us to see or experience. To uncover the "hidden" worlds of science all around us science writers slow things down, show the insides of things, or magnify things.

Zooming in, slowing things down, and considering how things work in a bit-by-bit way is of course exactly what students have already done in their narrative writing. As Annie Dillard observes in *Pilgrim at Tinker Creek* (2007), in our world, "There are lots of things to see, unwrapped gifts and free surprises." For science writers, and indeed for all writers, it is a matter of stopping to consider all there is to see and what they have experienced to write with greater insight and clarity, opening up new worlds for their reader.

IN THIS SESSION, you'll teach students that science writers use special strategies to share hard-to-understand concepts with their readers. Some of these strategies include slowing down the writing, magnifying pictures or images, and drawing pictures to show the insides of objects.

GETTTING READY

- ✔ Students' information booklets and writing tools
- ✔ Your teacher information book and your table of contents
- ✔ An example of a scientific discovery that stemmed from a scientist's ability to uncover the hidden story of a phenomenon (see Connection)
- ✔ A picture of the example of the scientific reference you make in the Connection; in our case, we show a picture of the peacock's curved, feathered tail to illustrate the observation that ultimately led the scientist to a discovery
- ✔ At least two examples of writing that are "hidden story" examples: one that is slowed down using a lot of steps, and another showing the inside of an object (see Teaching and Active Engagement)
- ✔ Your own forces and motion "hidden stories" for your information book chapters that you will use to model bringing out the "hidden story" (see Teaching and Active Engagement)
- ✔ Chart paper and marker to collect examples of partners' "Stars" and "Wishes" (see Share)

COMMON CORE STATE STANDARDS: W.2.2, W.2.5, W.3.2.a,b; RI.2.7, SL.2.1, SL.2.3, SL.2.4, L.2.1, L.2.2, L.2.3, L.2.6

MINILESSON

Showing Hidden Stories with Science Writing

CONNECTION

Offer writers a real-life example of the scientific process they have been working through.

"Writers, just last night, I read about a fascinating scientific discovery—let me tell it to you. A scientist named Angela Freeman was talking to a friend about peacocks. The friend looked at that gigantic feathered tail of the male bird and noticed something. Her friend said the tail fan seemed to be sort of curved inward, like a shallow satellite dish made for amplifying sounds! The two scientists thought about that and found it very interesting. Angela decided she would test to see if the male bird makes sounds beyond the usual sounds we always hear. If they did, she hypothesized that would help explain the shape of the peacock's tail!

"So, using a special machine that can record sounds below the human hearing range, Angela made detailed recordings of peacocks while they opened their tails. And guess what? Peacocks *are* making sounds that we can't hear when they spread their tail feathers—they make a low-pitched rumble called 'infrasound thrums.' Her hypothesis was right! When she played back the recordings to other peacocks, they responded with sounds humans could hear; they answered the recorded sounds as language!

"That experiment has given Angela the idea to test other birds to see if they also make sounds that are too low for human ears to hear. She thinks peacocks are probably not the only birds that do that. This discovery might open up a whole new set of understandings about how birds communicate.

"Angela noticed something in the world, made a hypothesis, tested it, and then from what she learned, she's making other hypotheses and experiments . . . just like you all have been doing in the past few weeks."

Explain that the job of those who study and write about science often is to explain things that are out of our range of perception—things that are hard for us to experience from regular life.

"What this article made me think about is this: the job of people who study and write about science is to explain things that we can't understand from just our usual senses. They explain things that happen too quickly or too slowly or are too small or too quiet for us to catch. People who study and write about science tell hidden stories!"

◆ COACHING

Teachers, this lesson has the potential to thrill and inspire your students because it addresses hands-on the excitement of using science to reveal new worlds—and to write about those ideas in ways that teach and inspire others. So don't hold back—share your own enthusiasm and curiosity with your students. They will follow your lead!

Name the teaching point.

"So writers, today I want to teach you that when people are writing about science—explaining things that are not part of everyday experiences—they use special strategies to show the hidden story of their topic. For example, they might slow things down, or show the insides of things."

TEACHING AND ACTIVE ENGAGEMENT

Explain and offer an example of slowing things down, writing lots of steps for one moment.

"Now I'm going to explain and offer you an example of each of these strategies. (After I explain each one, to see if you think it would help you as you write, you'll have a chance to figure out how you might use it in your own booklet.)

"You might decide you need to slow things down. Things may be happening so quickly in some situations that you need to say, 'Whoa! Let me tell you all the steps that are happening here, almost all at the same time. There is a hidden story here, hidden because it happens so quickly, and I need to tell it!'

"Here's an example of how someone writing about science might slow things down, using lots of steps. What if I wanted to write about getting an injury? In real life it might feel like: 'You get pricked by a thorn and you bleed.' If you are writing like a scientist, you tell the hidden story, slowing things w . . . a . . . y down. You would write a lot of steps, instead, something like this:"

1. With a thorn, you puncture the skin of your fingertip.
2. When you puncture the skin, you also puncture a tiny "pipe" that carries blood—a capillary.
3. In the wall of that capillary/pipe is "collagen." The puncture spills the collagen.
4. Blood sticks to the collagen and makes a scab.

"Wow, can you believe all that happens in those few seconds? There is a whole hidden story there, isn't there? And you can capture that hidden story by writing down a lot of steps for something that seems simple, filling them in bit by bit."

Ask students to find a spot in your table of contents where that strategy might help, then talk to a partner about how that part might go.

"Where do you think I could tell a slowed-down hidden story in my own information book? Will you help me look at my table of contents and imagine where I could put in some numbered steps, even where something seems simple?" I held up the chart paper that held my table of contents and paused while children contemplated this. "I hear a lot of ideas! I do agree that one spot could be my chapter 'Scrubbing Out Stains.' How might it go when I slow it way down with numbered steps? Turn and talk about how that part might go. I will think about that while you are talking."

As you demonstrate this strategy, you may want to emphasize how to search your writing for a part that will be tricky for your audience to understand. We want to carefully choose parts that will clarify information for our readers and aid comprehension of the text.

The notion of writing in small steps won't be new to students; they've learned to do this in narrative writing. However, they may need to do a little research to uncover "hidden" steps in their informational writing topics. If you choose to use our example of getting pricked by a thorn to model how to write with many steps, you might consider showing children a diagram of a capillary, or better yet, a picture of one in a book. Or choose another topic for which you have a visual aid or a book that clearly shows the steps you describe.

Demonstrate telling the slowed-down hidden story in one chapter of your topic.

"Okay, I'm going to try telling a slowed-down hidden story now about scrubbing out stains." I paused, thinking to myself, as I assumed a teaching and storytelling stance. "'You might think that scrubbing out a stain is as easy as rub, rub, rub, but did you ever think about what really has to happen for that to work? 1. When you wet the fabric, the water begins to loosen up the dirt. 2. Then, when you cover the stain with detergent it absorbs into the fabric and begins to emulsify, penetrating (or sinking into) the cloth and breaking up the stain. 3. Finally, as you quickly rub a scrub brush against the cloth, the friction loosens up the dirt further until it washes away.'

"Can you imagine telling a slowed-down hidden story in one of your chapters, describing lots of things that happen nearly all at once? Maybe you describe everything that happens when the cleat touches the soccer ball. Or what happens as you slam the door behind you. Turn and talk about how you might use this strategy!"

Explain and offer an example of showing the insides of something. Ask students to find a spot in their tables of contents where that strategy might help, and then talk to a partner about how that part might go.

"Before you resume your own writing work, I want to show you one more technique people who study and write about science often use—that is showing the insides of things. You all know the cross-sections books (Stephen Biesty and Richard Platt 1993, 1999, 2001). Those books, and lots of informational books, use drawings and labels to tell the hidden story about what is on the insides of things. Sometimes those are called 'cutaway diagrams.' Those kinds of pictures usually show part of the outside of things and then, a layer of what is under that outer layer that we normally can't see. Those inside parts are labeled with arrows to tell us how the parts move!

"Here is one example: this is a picture of a grandfather clock. It is turned at an angle in the drawing, so you can recognize it and see the front, but the side is drawn as though it is cut open so you can see into it. Then there are labels and arrows to show how the parts move: *pallet*, *escape wheel*, *weight*, and *pendulum*. Then, next to the labeled diagram is the explanation of how all those parts work together to make the clock work.

"Turn to your partner and tell him or her about a part of your information book where you could tell the hidden story by showing the insides of something to help your reader understand how it works. Like, 'What is inside a shoe or a soccer ball, and how does that help explain how it works? How do the parts move?' That sort of thing. Turn and talk."

LINK

Send writers off to apply, from now on, these or any other invented strategies to help them convey information about their topic, and forces and motion, to their readers.

"So, writers, whenever you are writing about science, like you are today, think about which of the ways you will use to reveal the hidden worlds that scientists study. Will it be by slowing down a moment step by step, one, two, three? Maybe by showing the covered-up insides of things and labeling the hidden parts with arrows to show how they move? Or some other way? Use all you know to reveal hidden worlds!"

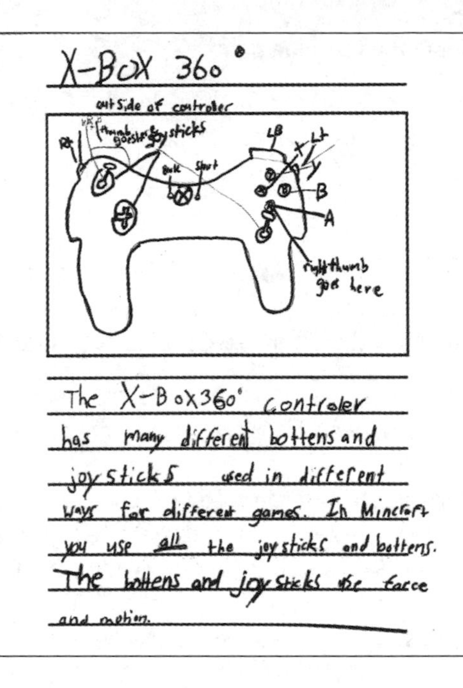

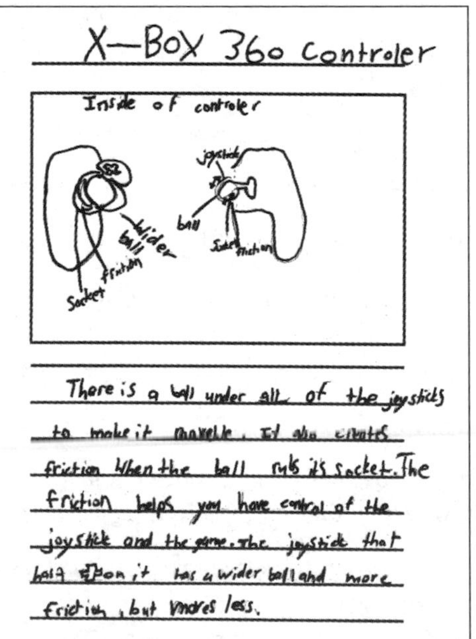

FIG. 11–1 Tyler shows the hidden story of an X-Box 360 controller.

CONFERRING AND SMALL-GROUP WORK

Working with a Partner to Envision and Act Out Parts to Imagine Smaller Steps

As you begin conferences today, you may notice that some students are attempting to show the hidden story behind their topics, but they need additional support. You may decide to pull a small group of students together and ask them to work in partnerships.

After convening the small group, quickly return to the example you offered in the day's minilesson. For example, you could say something like, "Remember when I wrote the hidden story behind scrubbing out stains? I didn't just say 'poof,' and it was down on paper. I had to do some work. I took a second to picture myself scrubbing out a stain, and then I actually pretended I was doing it. That really helped me remember the hidden story. This is something I thought we could practice with our partners."

Ask students to first close their eyes and picture the place where they want to show the hidden story. Then, ask Partner 1 to act out the part he would like to write bit by bit. Coach Partner 2 to look and listen carefully, naming out each small step Partner 1 mentions. Partner 2 might also ask clarifying questions to clear up any confusion. Partner 2 might say, "Can you stop for a second, rewind, and do that part again more slowly?" or "I'm confused with that part. Can you say and do it again?" After Partner 1 has acted out and written down the hidden story bit by bit, partners can switch roles. Then Partner 2 will have a chance to dramatize her work, while Partner 1 looks, listens, and asks clarifying questions.

On the other hand, if your students take to these ideas easily and are ready for more, you might offer up an additional strategy: creating a time lapse in their writing. Writers use time lapse the way photographers use it, to show in a short amount of time what happens over a long amount of time. Students who are writing about the friction involved in erosion, for example, might find this technique useful. They could describe a series of instants, hundreds of years apart. This technique is often used to show cause and effect or any relationship or change that we don't sit still long enough, or live long enough, to see.

MID-WORKSHOP TEACHING

Revealing Hidden Worlds by Magnification

"Writers, Joey found another way to help readers understand a hidden world. Let me share it with you in case you can use it, too! Joey found that it was easier for him to describe the friction that was happening between the surface of the ice and the surface of his skate if he drew a magnified, labeled sketch of the two surfaces touching each other, with some arrows to show how the surfaces rub together. Then he explained this in writing in a long column alongside the sketch. He was trying to explain why a sharpened skate can go farther and faster than an unsharpened one, and so he needed us to 'see' what we can't really see with the naked eye. So he drew a magnified version of it—he drew a small part bigger than it is in real life so we could see it more closely! Think quietly for a moment about whether magnification would help in any of your chapters." I paused a moment. "Even if it won't help you today, remember that is another technique that writers use, because it may help you someday."

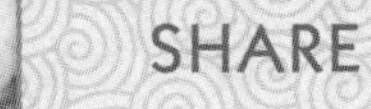

SHARE

Revising Based on Feedback

Set writers up to revise and accept feedback by giving a "star" and a "wish."

"Writers please join me in the meeting area. Bring your writing folder, and place it under your bottom so that we can use it in just a bit. All of your books now have several chapters, and they are coming along nicely. You have been thinking, writing, and revising on the run. When I write something really important, like the books you are writing, I ask a friend to read it and give me some feedback. I thought that would be helpful for you, too. We can give each other a 'star' and a 'wish.' When you give your partner a 'star,' you are going to give a compliment to his or her writing, and when you give a 'wish,' you are going to say something to your partner that you wish you had heard in the piece. Let me give you an example. My 'star' is that I really like how you used cross sections to help you make your book interesting to look at. My 'wish' is that you said more about the kinds of bikes that are good for riding in the mountains or on the beach.

"Okay, are you ready to try this with your partner? Partner 1, why don't you pick one page that you would like feedback on? Partner 2, your job is to listen carefully, thinking about the 'star' and 'wish' you would like to give to your partner. Partner 1, you can start reading." As students read their pages, I quickly listened in to several partnerships. I started to collect the kinds of "stars" and "wishes" students were saying. I started a quick T-chart with stars on one side and wishes on the other based on the feedback students gave each other.

Stars	Wishes
• Your pictures have labels and teach so much. • You are using different cross sections to make your writing easy to read. • You are using comparisons.	• I wish you said more about this part. . . . • I was a little confused with the last part. What did you mean?

Session 17

Introductions and Conclusions

Addressing an Audience

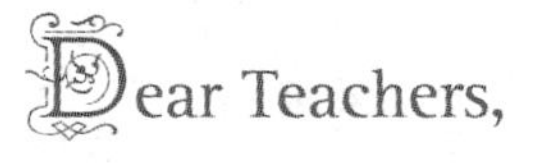
Dear Teachers,

People remember first impressions. Writers know this. You'll want your students to know this, too. In this session, you will work on introductions and conclusions with students. This process will deepen the work they did in first grade and will set them up for third-grade skills.

Last year, students learned to name a topic and provide some sense of closure, perhaps writing a thoughtful sentence or so at the end of their information books. Now, as second-graders, the Common Core State Standards expect students to "write informative/explanatory texts in which they introduce a topic, use facts and definitions to develop points, and provide a concluding statement or section" (W.2.2). You will want to see your students writing longer introductions and conclusions, elaborating on their ideas. Ideally, students should be writing several sentences for their conclusions.

You will also be encouraging your writers to be more aware of their audience. You might think of introductions as a drumroll for their information books—introductions set the tone for all of the teaching that follows. Likewise, conclusions should help readers understand the writer's aims by recapping important information and synthesizing main ideas. While the requirements for information writing at this level might stretch some students, remember that the format they will now use to set up and conclude their own information books is similar to that of so many books they pore over, daily, during their own reading times. This session again points students toward mentor texts with introductions and conclusions that engage and inform the reader.

MINILESSON

In the connection to your minilesson, you may decide to highlight to your students that in their day-to-day lives, they have devoted close attention to many kinds of introductions and conclusions. Ask them to recall the beginning or ending of a favorite movie, book, or

Common Core State Standards: W.2.2, W.2.5, W.3.2.a,d; W.3.5, RI.2.2, RI.2.6, SL.2.1, SL.2.3, SL.3.3, L.2.1, L.2.2, L.2.3, L.2.6

even a poem or song—if it's truly their favorite, they might even recall the exact lines with delight. This is because the writer of that text did his or her very best to make that introduction and conclusion memorable and powerful for the audience. Tell your students that today is the day they will do the same. Today is the day they will craft introductions and conclusions that are fun and engaging for their audience.

Draft a clear and explicit teaching point, such as, "Today I want to teach you that writers give their information books an introduction and a conclusion. When writing introductions and conclusions, writers try to get the reader's attention so they can highlight important information about a topic."

In your teaching, show students how you browse a few mentor texts to come up with ideas for how you might introduce your own teaching book. It will be helpful to have a few introduction and conclusion pages from your set of mentor texts already tabbed and ready to go. As you project these pages onto an overhead or document camera, remind students that they are looking at this with you to spark ideas for how they might engage their readers right from the start.

Sample Introductions

***Wolves* by Laura Marsh (from National Geographic Kids)**

What's That Sound?

Arroooooo!

There's a lonely howl in the distance. Then more voices join in. The chorus of howls sends a shiver down your spine.

What's making this spooky sound?

Wolves!

***Babysitting Basics: Caring for Kids* by Leah Browning (Capstone Press)**

Chapter 1: Yeah! The Babysitter's Here

Do you love children? Are you patient and responsible? Have you ever thought about becoming a babysitter?

As most babysitters will tell you, the job isn't easy. But it does have lots of rewards. Seeing a child's face light up when you walk into a room makes you feel great. And babysitting is a fun way to earn money after school and on weekends. You might not have experience with babysitting and aren't quite sure what to expect. Parents expect babysitters to be in charge while they're gone. This book will get you on the road to success.

You might point out to your writers that Laura Marsh introduces her topic, wolves, by posing a question. "What *is* that sound?" We read on because we want to find out! And Laura doesn't just give the answer away. She adds details to help readers imagine the setting, to draw them in, building suspense. Leah Browning, on the other hand, introduces her topic in a slightly different way. Leah also starts by asking questions that get readers thinking about her topic. But then, she lists reasons a reader might be interested in her topic and tells the reader what the purpose of her book is. Though these two examples offer possible starting places for your students, you may of course use other texts for your demonstration—like a class's favorite information book—depending on your writers' needs and interests.

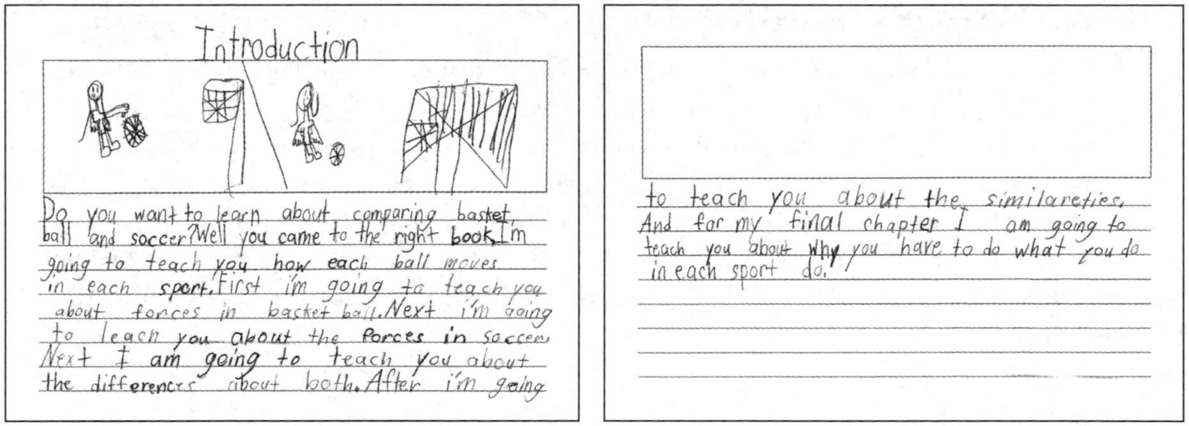

Introduction

Do you want to learn about comparing basket ball and soccer? Well you came to the right book. I'm going to teach you how each ball moves in each sport. First i'm going to teach you about forces in basket ball. Next i'm going to teach you about the forces in soccer. Next I am going to teach you about the differences about both. After i'm going to teach you about the similareties. And for my final chapter I am going to teach you about Why you have to do what you do in each sport do.

FIG. 17–1 Gabriella's introduction is a preview of her book.

For the active engagement, you might hand over the reins to your students, providing them with a range of books plucked from your classroom's nonfiction library baskets. These will be carefully selected books that will facilitate the work of mining for writerly gems. Your role will be to scaffold your students' inquiry into what the authors did that they too could try in their own introductions. It may be helpful to chart some of the observations your students discover. You might head the chart *Introductions Can . . .* and then bullet the possibilities: Pose a question, Put you in the place (setting), Start with dialogue or quotes, Ask a few crucial questions, Give a sneak peek, and so on.

CONFERRING AND SMALL-GROUP WORK

The next session will be the last day your writers have to tie up loose ends and flesh out parts of their writing that need extra attention. So, it will be helpful to be armed with a plan for small-group and strategy meetings. This means that prior to today you will have looked over students' booklets and your conference notes from the past week or so, reminding yourself of the course individual writers are on and identifying those who should get priority during independent writing time. In addition to conference notes and mentor texts, you will want to have available your own information book and student-sized copies of charts to coach writers on information writing strategies. Now's the time to encourage students to read aloud chapters to each other, listening for clarity and quickly fixing up writing that may not make sense, look right, or sound right. You may find that a handful of students could benefit from revisiting their tables of contents, revising to reflect newly added chapters as well as deleted ones. Some children may even decide to create new kinds of chapters, like glossaries.

MID-WORKSHOP TEACHING

In today's mid-workshop teaching, you might remind students that they learned how to conclude their lab reports in the first bends of this unit. You will want to make a distinction, however, between concluding a procedural lab report and concluding an information book. Students can study the conclusions of mentor texts in the same ways they've studied introductions, trying out different ways to write their own concluding sections. It may be helpful to revise your previous chart from Session 4 titled, "In Conclusions . . . " right on the spot, to illustrate the new work of concluding informational texts. You might replace old bullets with new ones: Say an idea and then say more about it, Recap the important parts, Congratulate readers, Send readers off, and so on.

Sample Conclusion

Babysitting Basics: Caring for Kids **by Leah Browning (Capstone Press)**

A Job Well Done!

Babysitting is an adventure. It can make you feel like a celebrity, a clown, or a zookeeper—sometimes all on the same night. But with the right preparation and patience, you can be ready for just about anything. So plan ahead, do your best, and enjoy the ride!

SHARE

To spotlight the journey your writers have taken in this unit, you may decide to build a bit of excitement around the science exhibit the class will have in the session after next. After all, your students have become deft writers of lab reports *and* information books! This is a huge accomplishment! Tomorrow will be a day to clear up and fancy up their writing. Starting tonight, students will need to give thought to the materials they want to bring in for the upcoming celebration, in which they will have a chance to present not only their writing, but perhaps some of the actual science work behind it as well. You might invite them to brainstorm ways to make their presentations interactive. Students could bring in photographs, images, or videos of their topics, or perhaps even the actual materials and equipment (for example, a hockey stick, puck, and shin pads) to demonstrate their topics!

Especially if students plan to bring in extra materials for presenting or demonstrating, you'll want to allot time for set up and practice before the celebration.

Good luck!

Lucy, Lauren, and Monique

Session 18

Editing

Aligning Expectations to the Common Core

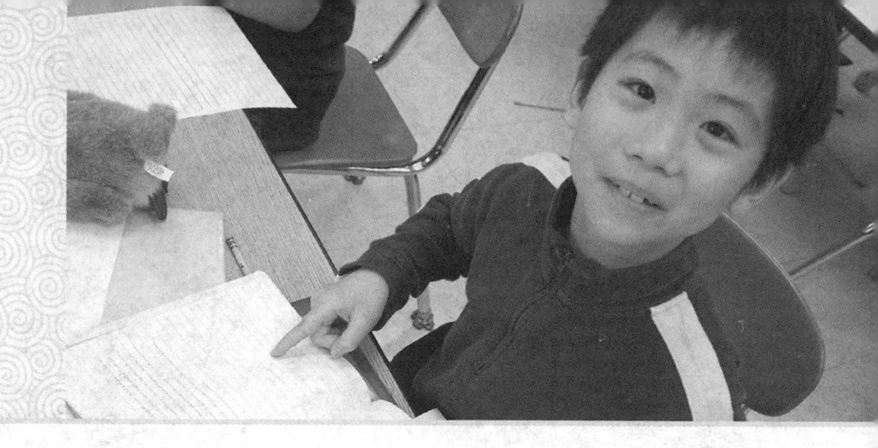

ENERGY WILL BE HIGH TODAY as students prepare to share and celebrate their hard work tomorrow. What a great opportunity to capitalize on that energy and focus it on fixing up their writing in ways that bring together all that they've learned throughout this unit. Today's work draws on the editing work embedded throughout this unit and on some of the editing strategies your students have learned in the previous unit, *Lessons from the Masters: Improving Narrative Writing*, as well as the second-grade foundational skills and language standards put forth by the Common Core State Standards. As you choose a focus for today's lesson, we also strongly suggest taking a close look through your students' writing. It is the one source that will help inform the valuable details of your teaching today.

You will teach into a repertoire of editing strategies students have accumulated over the past unit, from word study, and from the beginning of this second-grade year. In this session, you'll remind students of the editing work they've taken on with success. You may want to refer to the class word wall and word banks from this unit, and pull out past editing checklists and tools to help them recall and incorporate this work into their writing. They have carefully written with precise words, capitalized proper nouns, worked with various kinds of punctuation, and learned spelling patterns and strategies.

Along with a review of the work mentioned above, today's session presents you with the opportunity to introduce a few new editing strategies to add to students' growing repertoire. In preparation for this session, you may thumb through writers' booklets while keeping the Second Grade Common Core Standards for Foundational Skills and for Language close at hand. In the Foundation Skills (CCSS.RF.2.3) for second grade, you will find that your students are expected to know and apply grade-level phonics and word-analysis skills in decoding words, with examples ranging from distinguishing between long and short vowels to working with vowel teams and on to decoding common prefixes and suffixes. Furthermore, the second-grade Language standards (CCSS.L.2.2) require that students demonstrate command of the conventions of standard English capitalization, punctuation, and spelling when writing.

IN THIS SESSION, you'll teach students that writers edit their books by rereading and making their writing easier to read, inserting capitals, commas, and apostrophes where appropriate.

GETTING READY

- ✓ Extra copies of the Information Writing Checklist, Grades 2 and 3
- ✓ Enlarged class chart of the Information Writing Checklist, Grades 2 and 3 to use throughout the workshop
- ✓ Two chapters (or two pages) of teacher-created or student example of informational writing, written on chart paper, that can be edited for the conventions listed on the Information Writing Checklist, Grades 2 and 3 (see Teaching and Active Engagement)
- ✓ Note cards listing common contractions and possessives you see in your students' writing (see Conferring)
- ✓ Dry erase board and marker to demonstrate spelling a challenging word a few times (see Mid-Workshop Teaching)
- ✓ A few sets of simple dictionaries in the writing center for students to reference (see Mid-Workshop Teaching)
- ✓ Classroom computers set up for students to access Google.com as a spelling resource (see Mid-Workshop Teaching)

COMMON CORE STATE STANDARDS: W.2.2, W.2.5, W.3.5, RFS.2.3, RFS.2.4, SL.2.1, L.2.1, L.2.2, L.2.3, L.3.2.g

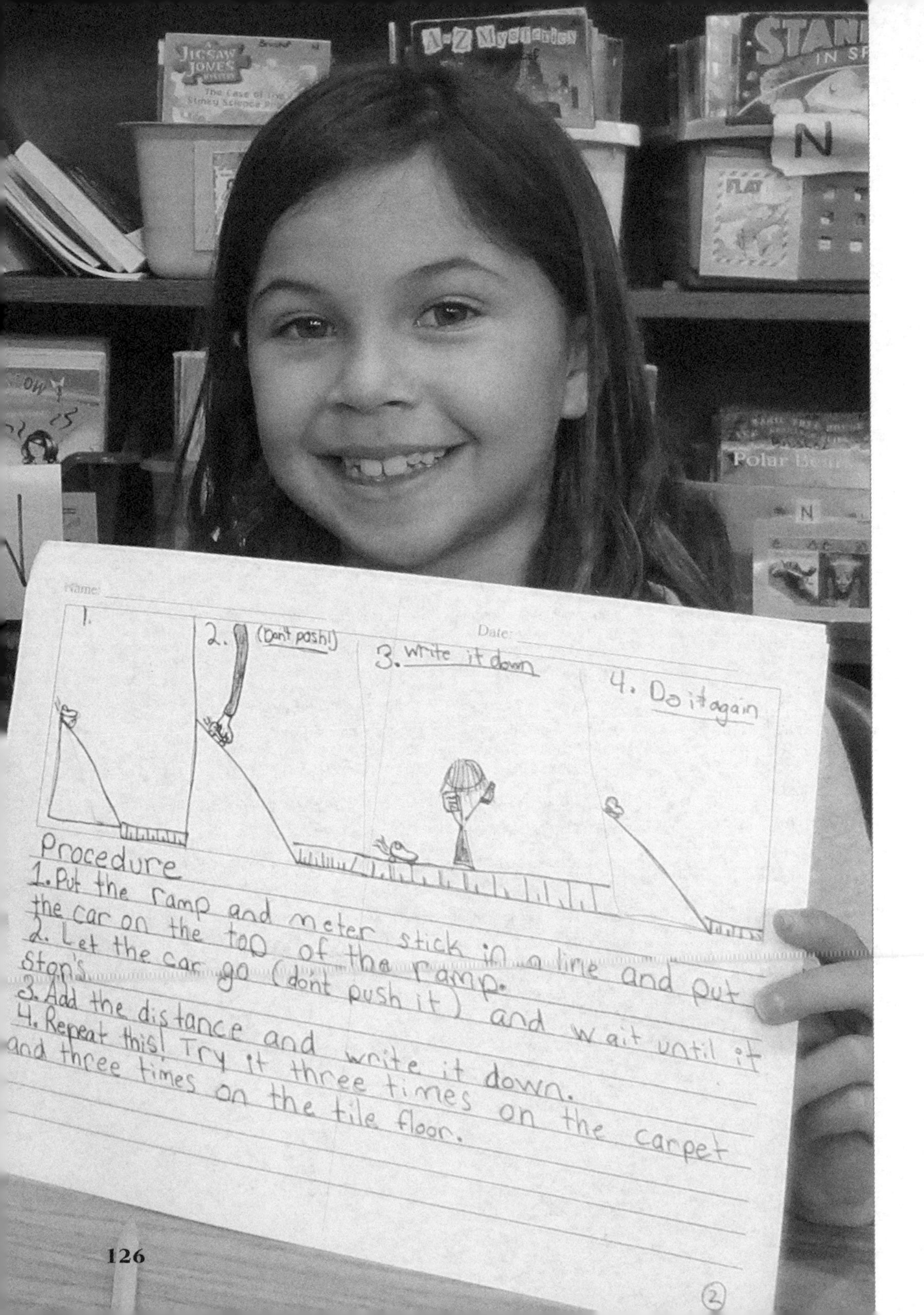

As you may have gathered, you will definitely have to prioritize, drawing on your students' writing to help you identify their common needs. This will help you determine what, specifically, to highlight in the minilesson, in the mid-workshop teaching, and in the share. We have selected to spotlight the work of editing for apostrophes, commas, and capitals because this work was a good fit with this class of writers. You will also notice that we tucked in a few new resources for checking tricky words. Beyond trying multiple ways to spell and fix up, we introduce dictionaries and/or computers as quick resources to help confirm the spelling of sophisticated words.

You will want to ensure that the mechanical work you teach mirrors students' abilities and approximations of what they know about conventions.

With this said, we advise that you not obsess about making these pieces conventionally perfect. Just as the science work reflects the science and information concepts each second-grade student has grasped, you will want to ensure that the mechanical work you teach mirrors students' abilities and approximations of what they know about conventions. This will provide you with valuable information as you plan your small-group editing lessons for the upcoming unit and the opportunity to show your students their growth over time in this part of the writing process.

Remember, this is a time to congratulate your students on applying their growing knowledge of conventions into their writing. This is no small feat! In the next session, they'll have their heads held high, marching into their celebration.

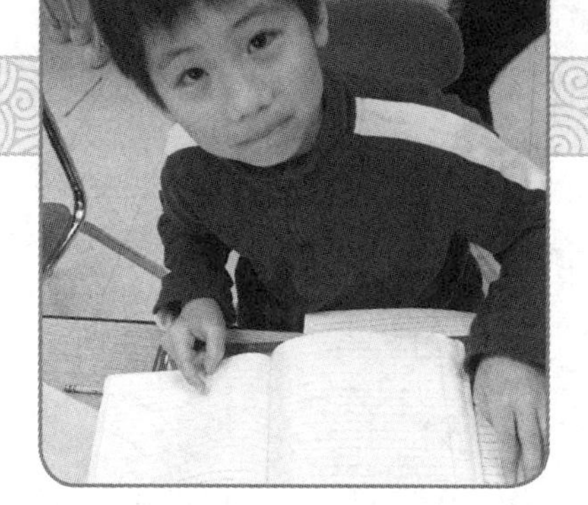

MINILESSON

Editing

Aligning Expectations to the Common Core

CONNECTION

Remind writers that they must edit their writing so that it is ready for readers at tomorrow's celebration.

"Writers, we are getting ready for our science exhibition celebration. At our celebration, you will share both your findings from your experiments and your information books with the world. We need to make sure everyone can read these easily! You have been working so hard to make your writing the best it can be by rereading and revising your work and by using the Information Writing Checklist to remind you of all that you need to do."

I pointed to the last part of our checklist as I said, "Now, before we celebrate, you need to make sure that your writing looks the best it can for your readers. Let's focus our attention on this last part of the checklist titled 'Development,' and 'Language Conventions.'" The Information Writing Checklist, Grades 2 and 3 can be found on the CD-ROM.

Information Writing Checklist

	Grade 2	NOT YET	STARTING TO	YES!	Grade 3	NOT YET	STARTING TO	YES!
	Structure				**Structure**			
Overall	I taught readers some important points about a subject.	☐	☐	☐	I taught readers information about a subject. I put in ideas, observations, and questions.	☐	☐	☐
Lead	I wrote a beginning in which I named a subject and tried to interest readers.	☐	☐	☐	I wrote a beginning in which I got readers ready to learn a lot of information about the subject.	☐	☐	☐
Transitions	I used words such as *and* and *also* to show I had more to say.	☐	☐	☐	I used words to show sequence such as *before, after, then,* and *later.* I also used words to show what didn't fit such as *however* and *but.*	☐	☐	☐
Ending	I wrote some sentences or a section at the end to wrap up my piece.	☐	☐	☐	I wrote an ending that drew conclusions, asked questions, or suggested ways readers might respond.	☐	☐	☐
Organization	My writing had different parts. Each part told different information about the topic.	☐	☐	☐	I grouped my information into parts. Each part was mostly about one thing that connected to my big topic.	☐	☐	☐
	Development				**Development**			
Elaboration	I used different kinds of information in my writing such as facts, definitions, details, steps, and tips.	☐	☐	☐	I wrote facts, definitions, details, and observations about my topic and explained some of them.	☐	☐	☐
Craft	I tried to include the words that showed I'm an expert on the topic.	☐	☐	☐	I chose expert words to teach readers a lot about the subject. I taught information in a way to interest readers. I may have used drawings, captions, or diagrams.	☐	☐	☐

Name the teaching point.

"Today I want to teach you to reread your writing with the lens of making it easier to read. As you reread, you can use the items on the Information Writing Checklist, Grades 2 and 3 to help you focus your attention."

TEACHING

Demonstrate using the checklist to edit a piece of writing.

"Writers, I have this piece of writing that needs a bit of fixing up. I am going to show you how I use the checklist to edit my writing. Watch closely because I am going to ask you to try this in just a bit.

"First, let me reread and think about the last section of the checklist. Then, I will reread my writing," I said as I turned my attention to the checklist.

"Okay, now let's look at this piece of writing to think about how I can edit it." I began to read:

Teachers, whatever writing sample you use here, you will want to be sure to stock it with those particular examples you want your students to practice here and to work on as they edit their own writing. Your sample may look quite different from ours, depending on our students' needs. Though of course we expect there will be some overlap of issues, most second-graders benefit from practicing.

Kinds of Bicycles

If your about to buy a bicycle, you need to know about all the different kinds out there. Some popular brands are trek cannondale specialized and raleigh.

The type of bike you buy depends of what you will want to use it for. Do you want it for the beach, mountains, racing, tricks, getting around the city? BMX bikes are great for doing all types of tricks or stunts. They are great for flipping around. Mountain bikes have thick tires and usually have great shocks for bumpy terrain. Racing bikes are made a light material and position your body to lean forward.

Highlight the use of *your* and change it into the contraction *you're*.

I read the first sentence aloud. "'If your about to buy a bicycle.' Wait, that doesn't look right. It is not that kind of *your*. 'If you are about to buy a bicycle' is the same as the contraction *you're*, with an apostrophe between the *u* and the *r*. You are. You're. I need to fix that." With my pen, I crossed out *your* and wrote *you're* on top.

Demonstrate using the checklist to edit for capitalization and commas.

"Let's read the next part. 'Some popular brands are trek cannondale specialized and raleigh.' Hmm . . . these are all brand names of bikes, right?" I looked back at the checklist, pointed to the item, and read, 'I used capital letters for names.' Those are not names of people; although, they are names of bikes. I have to change the first letter of each one to be a capital. You know, something else is not right," I said as I glanced back at the checklist. Students were lifting up their thumbs ready to come to my rescue. "I think I figured it out. I am listing the names of bikes. I need to add commas to remind my readers to pause between each bike name." I quickly inserted commas into the sentence and then continued to read the rest of the page.

Recap the demonstration, and name the teaching point with clear, concise language.

"Writers, did you see what I just did? Did you see how I studied the checklist, thinking about ways to make my writing easier for my readers to read? Then, I reread each part of my writing thinking about if there were places that needed to be edited. When I found something, I changed it! I changed *your* to *you're*." I pointed to the word and then slid my finger across the second sentence as I said, "Then, I noticed the names of the bikes needed to be capitalized, and I changed that. Finally, I added commas to the list of bikes."

ACTIVE ENGAGEMENT

Set writers up to practice this strategy on a shared text.

"Now it is your turn to try. I have another chapter from this book on bicycles that I thought we could study together. Let's think about what we need to do to make this writing easier to read." I gestured once again at the last section of the checklist. Then I flipped my chart paper to the next chapter and read aloud:

Hard-to-Ride Places

There are several places that make bike riding a bit more difficult. Some hard-to-ride places are sandy areas over rocky terrain and up hills. Thats just to name a few. Your tires may sink in sandy areas and its hard to ride. Rocky terrain produces tons of friction and it can be bumpy. Riding up hills can be hard because you have to do all the work pushing the bike uphill.

Give children a bit of wait time to apply the strategy.

"Many of you are already lifting up your thumbs! I'm guessing you found a bit that needs editing. Take a second to find not just *one* place but a *few* places tht could use editing." I gave children a bit of wait time to set the expectation that I wanted everyone to find a place to edit. When I noticed most of the children's thumbs raised I said, "Turn and talk to your partner about several places where we can make this piece of writing easier to read. Remember to use our checklist for help."

If students are struggling to find things that require editing in this piece, you might provide additional scaffolding and support. For example, you could ask students to study the piece through the lens of apostrophes. This additional support will help students focus on one particular thing.

Listen in to partnerships discuss editing strategies.

I listened in as partners described the work they would do.

"There is a list over there, where they name the hard-to-ride places. We need to add commas to show readers where to pause," Gabby said to her partner. Alexsa responded, "We could add a comma after 'sandy areas'."

As I moved to the next partnership, I heard Henry say, "The word *thats* doesn't look right." Sophia responded, "*Thats* is supposed to be *that's*. It's like *you're*. *Its* is also supposed to be *it's*."

Recap the work the students practiced.

"Writers, can I call you back for a minute? As I was listening in, I saw how you were wonderful detectives, picking out places where we can make this writing easier to read. You found a place to add commas and a place or two to add an apostrophe for words like *that's* and *it's*. Great work!"

LINK

Send students off to begin editing their own information books.

"Writers, as you go off today to get your writing ready to share with the world, keep in mind that writers always think about how to make their writing easy to read for readers. You can use the Information Writing Checklist to remind yourself of the strategies you are working on to make your writing easy to read. Then, you can reread your writing with this lens. Off you go!"

Supporting Writers' Usage of Apostrophes

IN PREPARATION FOR THE EDITING WORK your students will be doing today, you may want to create a strategy card listing contractions and possessives that are commonly found in your students' writing. It would be supportive to place copies of these cards in the writing center or to place one on each table for students to share.

During the minilesson, you discussed editing for the use of apostrophes. Some writers in your class were probably able to take the tip and apply it instantly. Other students may need more exposure and practice using apostrophes. You may inquire to find out if students understand apostrophe usage. If they need more support, you might lead a small group during which you give an explanation and example of how to use an apostrophe with both contractions and possessives. You may want to show a few words on a dry erase board not as contractions and then as contractions. You also may want to describe what the word *possessive* means and teach writers to ask themselves, "Does this object or thing belong to the noun—the person, place, or thing—in front of it? If so, you can add an apostrophe *s*." At this time, you also may want to give each student the strategy words containing frequently occurring possessives.

Then, you will want students to practice looking for words in their writing that require apostrophes. If, after this, students are still not using this editing strategy independently, they could practice rereading and editing in partnerships. Two sets of eyes are usually better than one! As partnerships reread, support your students by listening in and coaching them to notice more when they reread past something. You might say, "Why don't you reread that part again to see if you can find a place to edit?" "Check your strategy card to see if it can help you." This type of coaching still enables students to uncover, on their own, the work they need to do.

MID-WORKSHOP TEACHING **Using a Variety of Strategies to Spell Tricky Words Correctly**

I stood up in a central location in the classroom and said, "Writers, can I please have your eyes and ears?" I waited until all eyes were on me. "I love how you are working on rereading to edit your information books and want to give you a few tips to help you spell tricky words correctly. I know you know when a word isn't spelled right, but sometimes you just leave it because you don't know what to do. Well, here are a few tips to help you spell the words correctly. One thing you can do is to write the word a few times, trying out different ways you think it is spelled. Then, you can circle the one that looks best and check it with you writing partner. For example, if I wrote *reseve*," I wrote the words on the white board as I said them, "I could try it out like this: receve, recieve, receive."

I moved in front of the writing center, grabbed a dictionary, and said, "Another strategy you could try is to use one of our dictionaries to find the spelling of the tricky word and fix it if necessary."

Then I walked near our classroom computers, gestured toward them, and said, "A third strategy you could try for spelling a tricky word correctly is to Google it. You can go to Google.com and type the word in the search box and hit enter. When the search results come up, it might say, 'Did you mean . . . ?' Usually, the correct spelling of the word you are looking for comes up.

"Writers, you know so many ways to figure out how to spell these tricky words. It's your job to fix them!"

SHARE

Reflecting on the Second- and Third-Grade Information Writing Checklist

Recruit writers to assess their writing against the second- and third-grade Information Writing Checklist, in preparation for tomorrow's celebration.

"Writers, please join me at the meeting area with your writing." I turned the easel back to face the meeting area. "Throughout this unit, you have been using your checklist to remind yourselves of what your writing needs. It has been a great tool to keep you on track. Look at this checklist one more time before our celebration to pick which item you feel you most improved on and are proud of. I'll give you a bit of time to look between the checklist and your writing."

As students turned to this task, I looked at them to get a sense of their progress. "I know it is hard to pick just one example but choose one and place your finger next to that part. When I pass my imaginary baton, please name out which item on the checklist you picked and show or read the example. Let me get the baton." I reached behind the easel to grab the imaginary baton.

I pointed it to Sophia who said, "I used a diagram and a table in my writing."

"Oh, so, Sophia, you really worked on including facts and details in your writing!" I said as I pointed to that section of the checklist.

I pointed the baton at Tyler. "I really worked on my capital letters. I had a lot of names of things like you did today, and I didn't put a capital letter. So, I went back to my writing and edited it. I was writing about ice hockey teams, and I capitalized the Rangers and Devils."

I pointed the baton a few more times to give other students a chance to reflect on how they had grown as writers. "Writers, I can't wait for our celebration tomorrow. We are ready to share our writing with the world!"

Session 19

Celebration

Writing and Science Exhibition

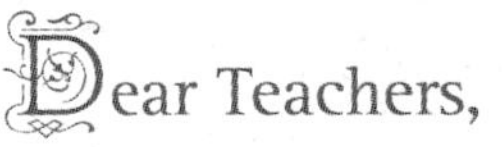
Dear Teachers,

You have lots to celebrate. In fact, today is a doubleheader celebration! Today is the day to bask in the maturation of your students as information writers and to showcase their knowledge of the new scientific concepts they have come to master.

Your students will never again be the same kind of information writers they were before you began this unit. You've opened their eyes to the world of science studies, specifically to the physical science concepts of forces and motion, and they've learned that informational writing is much more sophisticated than simply writing facts about a topic of expertise. They now know that writers can weave in knowledge learned from other content areas to help them think, write, and teach things they know a lot about, through a whole new lens, a scientific lens.

It is important to think about the huge journey your writers have taken and the kind of celebration that this warrants. We recommend that you pull out all the stops and reserve a vast space in your school building, say, the gymnasium or a multipurpose room, to set up a science exhibition that will be interactive and hands-on for your visitors. Rather than invite one particular class, you may decide to choose one period in the day to open the event to any member of the school community who will be available at that time.

Your students can create signs advertising the day and specific time of the exhibit, making sure that everyone in the community will be able to know of the celebration. You may have students post some flyers by the school entrances, so the crossing guards, visitors, and parents will be in the know. They can also hang signs around the lunchroom so that the fellow schoolmates, school chefs, and the school's maintenance workers may visit. Hand-deliver flyers to the especially curious classes that may have wondered about the ramps they saw set up in the hallways weeks ago.

Your guests will take part in the simple experiments your students designed from the beginning of the unit and will read their corresponding lab reports and information books to see the science behind the movements people perform each and every day. The visitors

Common Core State Standards: W.2.2, W.2.6, W.3.8, W.3.10, RFS.2.4, SL.2.1, SL.2.3, SL.2.5, SL.2.6, SL.3.5, L.2.1, L.2.2, L.2.3

will leave the celebration feeling like scientists. They will have the opportunity to do the work scientists do—reading experiments, wondering, and trying them out. Students can mark important pages in their information books that they want to be sure to highlight when visitors stop by their booths. When possible, students can bring in props to support the teaching they are doing in their information books. For example, participants could try to push the hockey puck across the surface.

After the exhibit, if you have the time and energy you could continue the celebration classroom, challenging students to create and invent purposeful, long-lasting teaching artifacts to make for those who were unable to visit the exhibit. For example, students could collaborate in small groups to design simple games teach others about their topics or about the science behind their topics.

PREPARATION

Prior to the celebration, students can come in wearing "lab coats" (oversized men's shirts) and set up a thematic science exhibition in the school gym: they can set up their books, materials/props, and other media (laptops, iPads set up to display videos taken, photographs, pictures, images printed from the Internet) and maybe bring back some of the cool experiments from the beginning of the unit. They could even make new kinds of experiments or mini replicas of a motion that relates to their information book topics.

CELEBRATION

During the celebration, students will man their exhibit stations while guests come in to read lab reports and do some of the hands-on experiments. Students can teach about friction, movement, force, and their topics, perhaps demonstrating with some of the materials or props they have to illustrate the concepts behind the information they teach. They can quickly reference the marked pages of their information books to read aloud or show.

AFTER THE CELEBRATION

Following the celebration, you may decide to keep this work alive in your class. If so, tell students that you wouldn't dream of packing away all this knowledge and information into file folders, never to be thought of again. Back in the classroom, they can design forces and motion board games in small groups. You may wish to print out sample board game templates (there are several ones in lots of word study or mathematics books). If you wish to go all out, you can purchase sturdier ones from teacher or craft stores. Students can quickly brainstorm their game name and concept ("Chutes and Ladders" = "Catapults and Ramps"), write up directions (procedures) for how to play the game, design game pieces (like the cotton balls, cars, snap cubes), and write up challenge questions and answers on game cards (Q: Where's the best place to ride a scooter? A: On a smooth surface.). These games can live in the classroom for them to play (and add

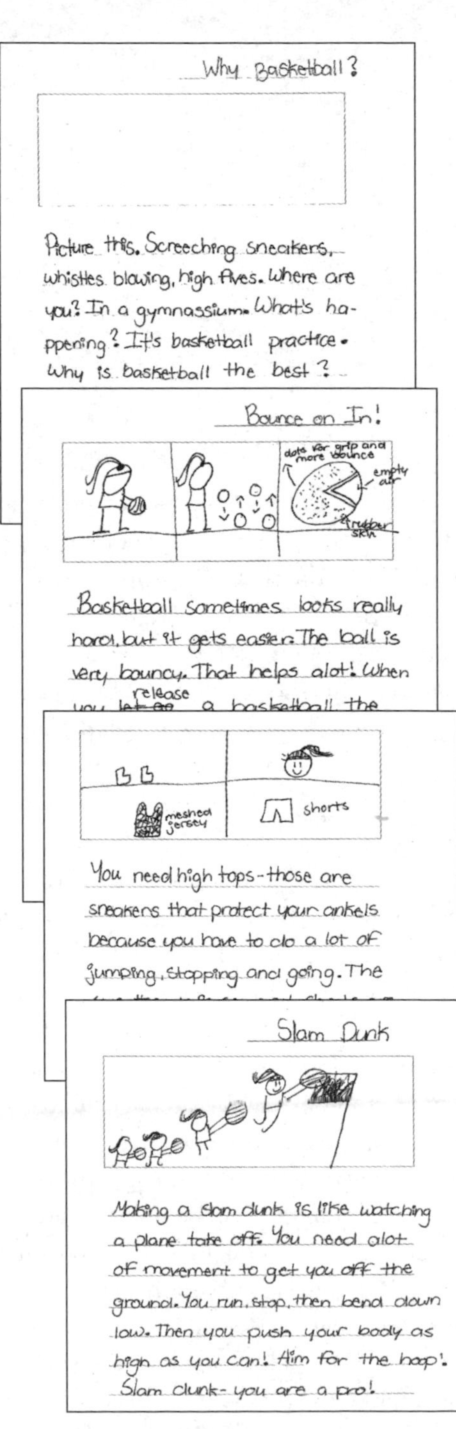

FIG. 19–1 Sisi's finished information book

on to) during choice time or be given as a borrowed gift to some of the classes that attended their exhibition or that were unable to make it. The games could even circulate around the school to several classes, spreading the information and joy of learning to write about science and information.

Enjoy!

Lucy, Lauren, and Monique

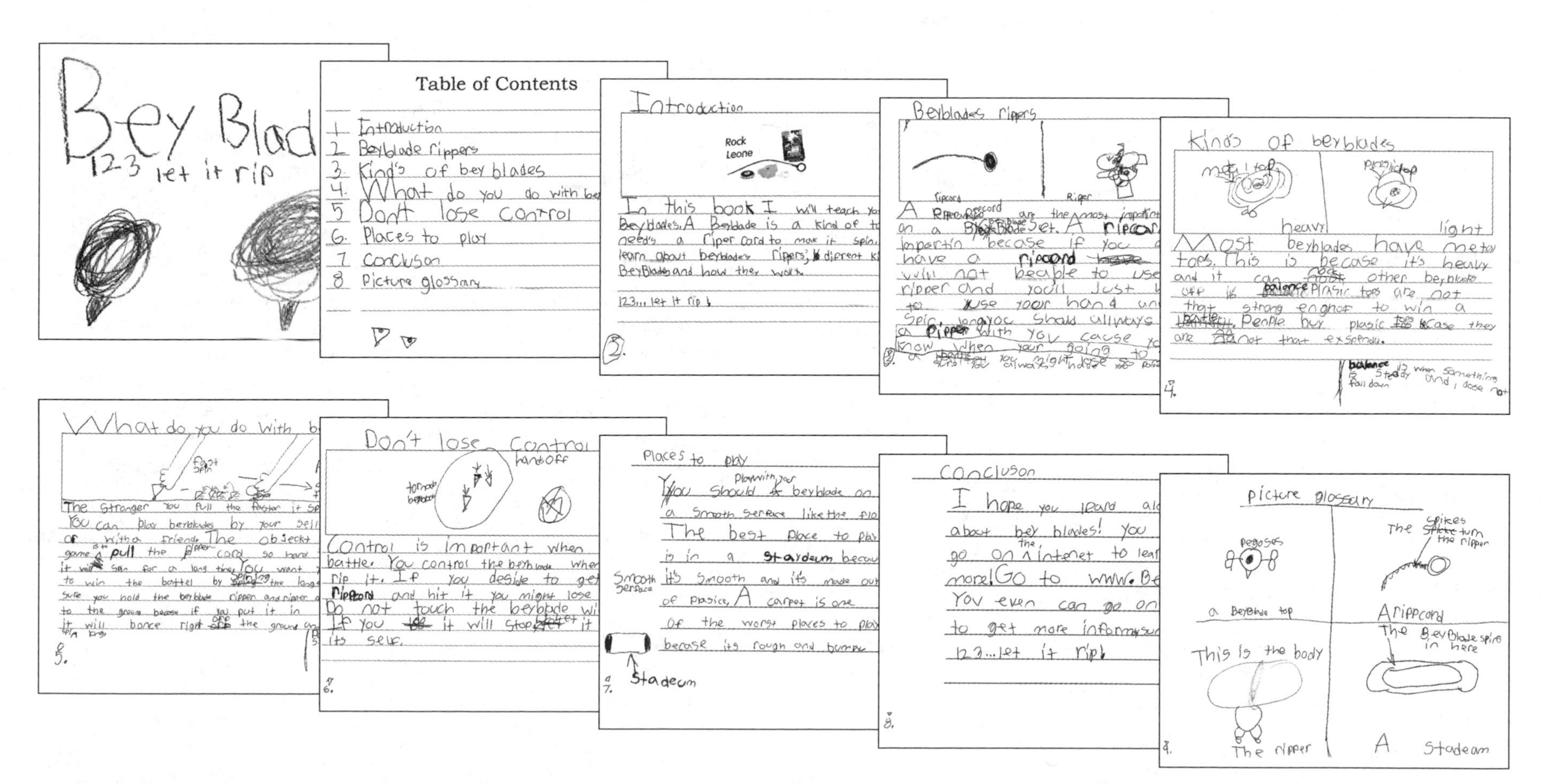

FIG. 19–2 Bey Blades info book

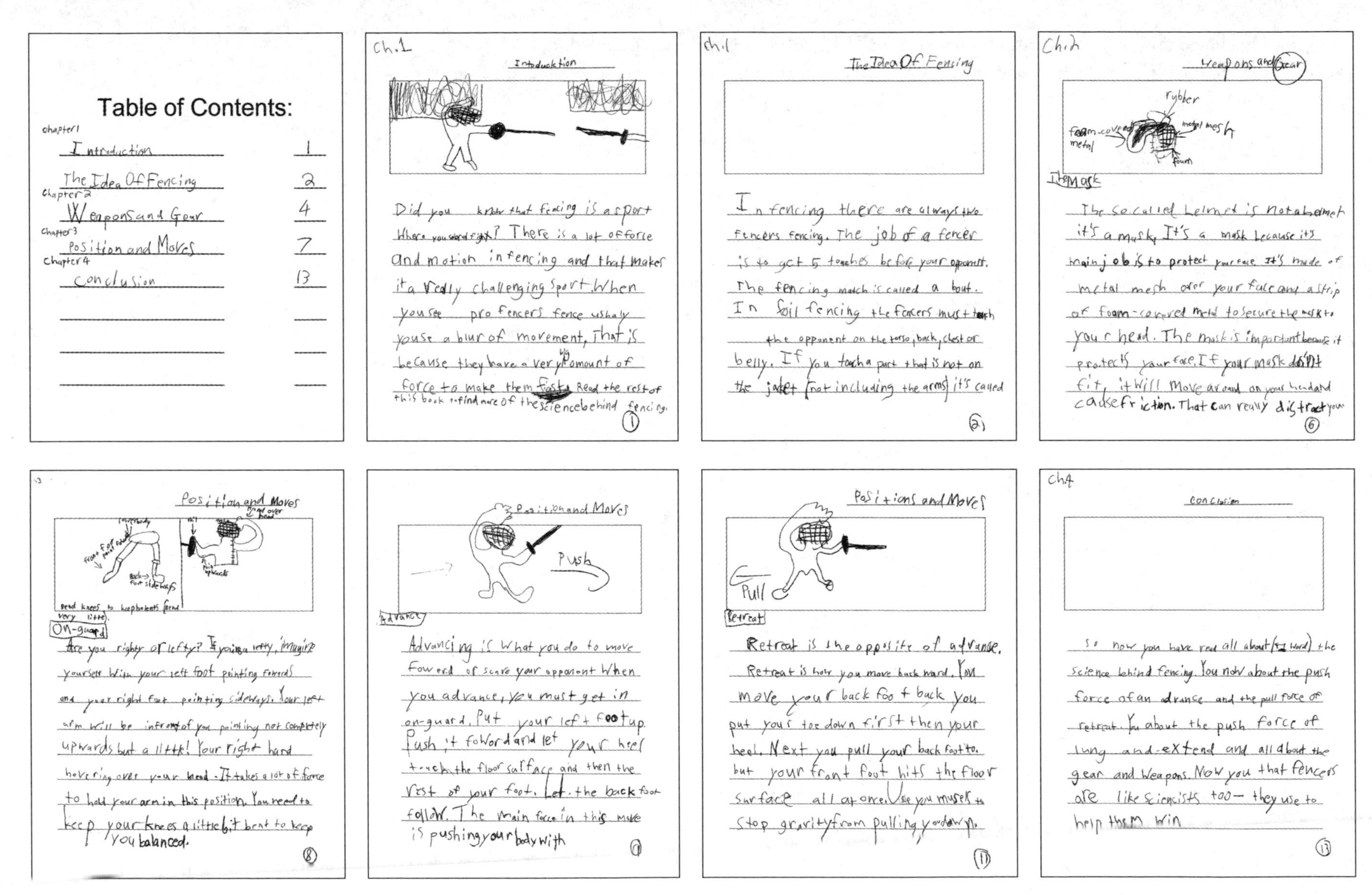

Table of Contents:

chapter1
Introduction 1
The Idea Of Fencing 2
chapter2
Weapons and Gear 4
chapter3
Position and Moves 7
chapter4
conclusion 13

Ch.1
Intoduckfion
Did you know that fencing is a sport where you sword fight? There is a lot of force and motion in fencing and that makes it a really challenging sport. When you see pro fencers fence ushaly youse a blur of movement, That is because they have a very big amount of force to make them fast. Read the rest of this book to find more of the science behind fencing.
1

ch.1
The Idea Of Fencing
In fencing there are always two fencers fencing. The job of a fencer is to get 5 touches before your opponent. The fencing match is called a bout. In foil fencing the fencers must touch the opponent on the torso, back, chest or belly. If you touch a part that is not on the jacket (not including the arms) it's called
2

Ch.2
Weapons and Gear
rubber
foam-covered metal
metal mesh
foam
The Mask
The so called helmet is not a helmet it's a mask. It's a mask because it's main job is to protect your face. It's made of metal mesh over your face and a strip of foam-covered metal to secure the mask to your head. The mask is important because it protects your face. If your mask doesn't fit, it will move around on your head and cause friction. That can really distract you
6

ch.3
Position and Moves
hand over head
lower body
front foot point forward
back foot sideways
point upwards
bend knees to keep balance (bend very little).
On-guard
Are you righty or lefty? If you're a lefty, imagine yourself with your left foot pointing forwards and your right foot pointing sideways. Your left arm will be in front of you pointing not completely upwards but a little! Your right hand hovering over your head. It takes a lot of force to hold your arm in this position. You need to keep your knees a little bit bent to keep you balanced.
8

Position and Moves
Push
Advance
Advancing is what you do to move foward or scare your opponent. When you advance, you must get in on-guard. Put your left foot up. Push it foword and let your heel touch the floor surface and then the rest of your foot. Let the back foot follow. The main force in this move is pushing your body with
9

Positions and Moves
Pull
Retreat
Retreat is the opposite of advance. Retreat is how you move backward. You move your back foot back you put your toe down first then your heel. Next you pull your back foot to. but your front foot hits the floor surface all at once. Use you musels to stop gravity from pulling you down.
11

Ch.4
conclusion
So now you have read all about (1 word) the science behind fencing. You now about the push force of an advance and the pull force of retreat. You about the push force of lung and extend and all about the gear and weapons. Now you that fencers are like sciencists too — they use to help them win
13

FIG. 19–3 Judah's book on fencing